Specifications for the Chemical and Process Industries

Also available from ASQC Quality Press

ISO 9000 Guidelines for the Chemical and Process Industries, Second Edition
ASQC Chemical and Process Industries Division,
Chemical Interest Committee

Quality Assurance for the Chemical and Process Industries: A Manual of Good Practices
ASQC Chemical and Process Industries Division,
Chemical Interest Committee

Total Quality Management in the Clinical Laboratory
Doug Hutchison

Guidelines for Laboratory Quality Auditing
Donald C. Singer and Ronald P. Upton

To request a complimentary catalog of publications, call 800-248-1946.

Specifications for the Chemical and Process Industries

A Manual for Development and Use

ASQC Chemical and Process Industries Division

Chemical Interest Committee

ASQC Quality Press
Milwaukee, Wisconsin

Specifications for the Chemical and Process Industries:
A Manual for Development and Use
ASQC Chemical and Process Industries Division, Chemical Interest Committee

Library of Congress Cataloging-in-Publication Data

Specifications for the chemical and process industries: a manual for development and use / ASQC Chemical and Process Industries Division, Chemical Interest Committee.
p. cm.
Includes bibliographical references and index.
ISBN 0-87389-351-4
1. Chemical industry—Quality control. I. American Society for Quality Control. Chemical Interest Committee.
TP149.S655 1996
660'.068'5—dc20 95-45500
CIP

10 9 8 7 6 5 4 3 2 1

ISBN 0-87389-351-4

Acquisitions Editor: Susan Westergard
Project Editor: Kelley Cardinal

ASQC Mission: To facilitate continuous improvement and increase customer satisfaction by identifying, communicating, and promoting the use of quality principles, concepts, and technologies; and thereby be recognized throughout the world as the leading authority on, and champion for, quality.

Attention: Schools and Corporations
ASQC Quality Press books, audiotapes, videotapes, and software are available at quantity discounts with bulk purchases for business, educational, or instructional use. For information, please contact ASQC Quality Press at 800-248-1946, or write to ASQC Quality Press, P.O. Box 3005, Milwaukee, WI 53201-3005.

For a free copy of the ASQC Quality Press Publications Catalog, including ASQC membership information, call 800-248-1946.

Printed in the United States of America

Printed on acid-free recycled paper

ASQC
Quality Press
611 East Wisconsin Avenue
Milwaukee, Wisconsin 53202

Contents

Preface

The Chemical and Process Industries Division of the American Society for Quality Control (ASQC) is committed to quality improvement throughout the chemical and process industries (CPI). Since 1984 the Chemical Interest Committee (CIC) has worked to reach consensus and to document good quality practices for the CPI. In 1987, through the auspices of Quality Press, the CIC published *Quality Assurance for the Chemical and Process Industries: A Manual of Good Practices.* Affectionately referred to as the "little red book," it has been a best-seller for ASQC Quality Press.

When the ISO 9000 standard for quality systems was first released in 1987, the CIC saw a need to link the concepts of good practices in its publication with the new standard. In 1992 the CIC published an ISO 9000 guidelines book. This was also a best-seller for ASQC Quality Press, and the "little blue book," *ISO 9000 Guidelines for the Chemical and Process Industries, Second Edition* has now been released.

This new volume on specifications marks a further step in the process of clarifying issues and seeking consensus practices for the CPI.

Acknowledgments

The Chemical Interest Committee of the Chemical and Process Industries Division of the American Society for Quality Control has prepared this manual with the intent of clarifying the setting and interpretation of specifications in the chemical and process industries. The CIC members who participated in the process of writing and editing are

Mark Berenson
Jim Bigelow
Candy Bolles
Bradford Brown (Chair)
Georgia Kay Carter
Ken Chatto
D. C. Cobb
Jeffrey Dann
Patrick P. Donnelly
Don Engelstad
David Files
Richard Hoff
Rudy Kittlitz
Norman Knowlden
Percival Ness
William Ochs
Chris Ostman
Frank Sinibaldi
Ray Vlasak
Jack Weiler

The support of the following companies is hereby acknowledged.

AlliedSignal Inc.
Bradford S. Brown, Consultant
Champion International Corporation
Ken Chatto, Consultant
Drexel University
DuPont
Eastman Kodak Company
Exxon Chemical Company
Hercules Incorporated
Occidental Chemical Corporation
OSRAM SYLVANIA INC.
Omni Tech International Ltd.
Shell Chemical Company
3M Company
Quantum Chemical Company
SCM Chemicals
Witco Corporation

Introduction

This book presents an overview of practical information about specifications in the chemical and process industries (CPI). The intended readers are individuals involved in quality, engineering, manufacturing, sales, marketing, and purchasing within the supplier, producer, and customer organizations. Readers may use this book as a guide and a reference for establishing, negotiating, and using specifications. Workers in other industries will likely find many of the principles and concepts in this book applicable to their work with specifications.

The most important element in the business of manufacturing is the transaction in which a customer purchases a producer's goods and services. If the transaction is mutually beneficial, each party's set of expectations is met. Although there are other factors, the producer is satisfied if the customer pays in full for the product, while the customer is satisfied if the product is fit for use and meets expectations.

In order for the producer to increase the probability of future business with the customer, a dialogue between the two is initiated. The producer learns of the customer's requirements and expectations; the customer learns of the producer's capabilities and limitations. The outcome of this dialogue is the agreed requirements for the product that are recorded in a document called a *specification.* This specification serves as the reference for the acceptability of the product in future transactions.

The specification documents the business journey jointly embarked upon by a producer and customer. They are each continually changing and improving their products and technologies, which places different pressures and expectations on the product. Specification requirements are rarely permanent, and they need periodic review to ensure their continued adequacy.

Specifications are the interface between the realities of the producer's process and the needs of the customer's process. Even if the customer does not require a specification from the producer, as with many consumer goods, it is still good practice for the producer to manufacture products to meet a specification. It has proven to ensure the consistency of product over time.

The CPI routinely make use of specifications. There are many characteristics and issues that may be unique to the CPI. Many of these were identified in *Quality Assurance for the Chemical and Process Industries: A Manual of Good Practices* (1987, 1–2).

> The art and science of sampling and measurement are different in chemical and process industries compared to mechanical industries. The relatively large contribution of measurement error requires proper use of statistical methods for process control. Because of this, proper interchange and understanding of information between customers and suppliers becomes very important.
>
> Additional considerations requiring special attention and a different framework to apply statistical methods in the chemical and process industries are listed as follows:
>
> - Raw materials are often natural materials whose consistency depends on natural forces. This requires making compensating adjustment in the process to maintain a consistent quality level of the finished product.
> - The technology of chemical sampling presents complex problems because of chemical considerations and the varied nature of product units ranging from a few grams to thousands of tons.

- Chemical processes occur on the molecular level and changes are often inferred from secondary information.
- Measurement itself is a complex process which must be carefully standardized and controlled.
- Automatic process control is highly developed with computerized feedback and feed-forward loops. Emphasis is on "prevention."
- Once produced, product properties often change with time as well as from changes in environmental conditions.
- There may not be a direct relationship between the measured properties and the performance characteristics of the products in use.
- True defectives are rarely found, although non-uniformity between lots is often detected by the customer.
- Health, safety, and environmental impact of both the process and the product are major concerns.

Specification values are numerical or verbal expressions of customer needs. Specification values should be the result of negotiations and should have a basis in reality. The ability to meet defined specifications has become a requirement of rational, economic competition. Setting limits that are more restrictive than the process capability invites failure on the part of the producer and dissatisfaction on the part of the customer. Sales, marketing, and purchasing personnel often are on the front line of the negotiation.

As with any business transaction, there are costs and risks associated with the customer and producer mutually agreeing upon a specification. It is not uncommon that a product can meet all of the requirements defined in the specification and yet be unsuitable. Specifications are plans that, although good to follow, can never fully define all unacceptable characteristics. In addition, the customer and producer must acknowledge that the specification content can be as dynamic as the sum of the changes made by both parties to their respective products and manufacturing operations.

The development and maintenance of a specification is truly an iterative process whereby an ongoing dialogue between the customer and producer can define, not only minimum requirements, but also needs and wants. In this way, the specifications provide benefits in helping achieve the following strategic business objectives for both producer and customer.

- To guard against economic loss at the individual purchase level (mitigation of complaints, disruption of production, and returns)
- To establish a formal means to convey product and process changes and their implications
- To define the relative importance of the defined characteristics for improvement plans, thus aligning the efforts of the producer to achieve changes most desired by the customer

It is the consensus of the authors that all specifications, both internal and external, share common elements. We believe that an understanding of the common elements will make working with specifications easy and productive. We have assembled a process for setting specifications that we recommend for use. We believe that using this process will civilize, if not completely resolve, the inevitable conflicts of perspective and expectation of those involved in the negotiation of specifications.

Philosophy of Specifications

In the three-year process of preparing this book, the authors have come to a consensus on the elements of a philosophy for specifications. It includes six principles. These principles are briefly stated below and developed in the chapters referenced.

- All specifications contain common elements (chapter 3).
- All requirements should be clearly documented and preserved in a controlled document system (chapters 5 and 21).
- The targets for product characteristics should be established before limits are discussed (chapters 7 and 8).
- Specifications are the result of negotiation between the producer and the customer (Introduction, chapters 1 and 20).
- The agreed requirements should be explicit, not implied (chapters 5 and 21).
- Data should be used as the basis for negotiations (chapters 10–15).

Part I

The Specification System

CHAPTER 1

The Specification System

1.1 The Supply Chain

ANSI/ASQC A3-1987 gives the following definition of a specification: "The document that prescribes the requirements with which a product or service has to conform." A specification is a basic component in a contract between a producer and customer. As with contracts, specifications may be established between governments, companies, departments, groups, or individuals.

In the chemical and process industries (CPI), we often speak of a *supply chain*, in which a producer buys something from a supplier, treats or converts it in some way, and then sells it to a customer. Throughout this book, we will speak of the *supplier*, the *producer*, and the *customer*. We have written the book and designed the specification system from the producer's viewpoint.

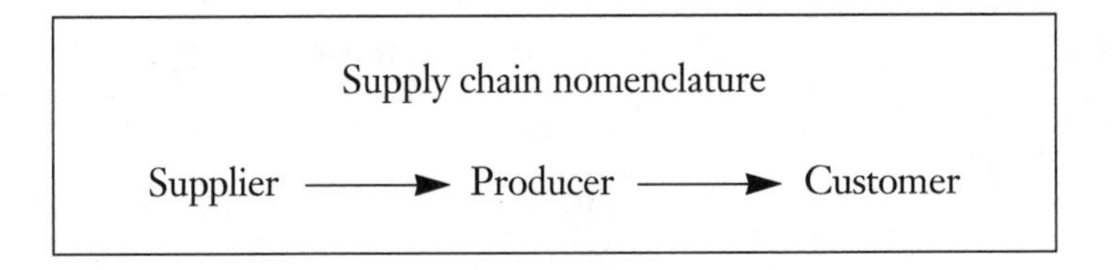

1.2 Book Outline

This book is divided into six parts.

- Parts I, II, and III progress from a description of the specification process, to a discussion of the components of the specification, to the details of the data required to decide upon the values for the specification targets and limits.
- Part IV discusses issues involved in the use of the specification created in Parts I, II, and III.
- Part V considers moving toward improved relationships with suppliers and customers and the issue of continuous improvement.
- Part VI gives a case study for a hypothetical product "ThanQs" that has many elements found in typical CPI processes.

1.3 The Process for Developing Specifications

Developing a specification can be performed in a five-step process as depicted in Figure 1.1. This process is readily applied to external specifications requests (for example, the development of a customer specification) or to internal specifications requests (for example, the development of a synthesis specification). Each of the five phases identified in the figure is discussed in the following text. It is assumed that the producer has delegated the responsibility to manage that particular type of specification to an individual or organization.

Phase I	**Initiate the process.**
Phase II	**Manage the process.**
Phase III	**Document the specification.**
Phase IV	**Approve the specification.**
Phase V	**Distribute and maintain the specification.**

Figure 1.1. Specification development process.

1.3.1 Phase I Initiate the Process

The specification development process is customer- or market-driven. It is also seen as a responsive process for which the trigger is a request to either develop a new specification or revise an existing one. Another input may be the obsoleting or deleting of a specification.

Specification requests should be recorded. This ensures that the required minimum information is received to plan and execute the process. Pertinent information includes

- Date received
- Requester
- Product
- Internal contacts
- Who will handle the request
- Associated correspondence
- Urgency
- Date for completion

1.3.2 Phase II Manage the Process

There are basically three types of specification requests. The process used to handle each type will differ.

1. Revising a specification implies that only a portion of an agreement is subject to change and that the issues surrounding the revision are understood. In revising an existing specification, the impact of a proposed change must be assessed before a decision can be made to make the change. If the proposed change calls for the addition or deletion of a characteristic, the result will be a new specification.

2. Establishing a new specification implies that the draft used to negotiate the requirements will be developed by the producer. This is advantageous to the producer because the requirements are best understood within the producer's organization.

3. Responding to a customer's specification is the most complicated process. More work is required on the producer's part to respond to a customer-generated specification because each requirement must be reviewed for implicit meaning.

Once the type of request is known, planning for its successful completion requires an understanding of the following:

- *Customer issues:* In order to end up with a meaningful specification, a thorough understanding of the customer's requirements and all the issues surrounding the request is needed, most importantly, why the request was made (for example, in response to a complaint). Another issue may be a requested completion date, such as a customer product specification requiring review within 48 hours.

- *Producer capabilities:* The producer's capabilities must then be compared with the customer's requirements to determine if they can be met. In addition, evaluate the resources needed to satisfy the request.

- *Conflict resolution:* Knowledge of approval and conflict resolution is required for those instances where the producer's capabilities cannot meet customer requirements.

1.3.3 Phase III Document the Specification

Once all of the issues are understood, the challenge now becomes one of documenting them in a manner acceptable to all involved. This includes defining the following:

- The characteristics that determine achievement of agreed-upon requirements
- The test or inspection methods (measurements) used to evaluate the characteristics
- The targets and limits that define acceptability
- The sampling method

Specifications for existing products can be used as templates for new, similar products. It is important to ensure that all the characteristics of the existing product are appropriate for the new product, that additional characteristics are not needed, and that the test methods and sampling plan are valid.

The data and logic used in making the choices recorded above should be retained for the life of the specification because of the time that may be saved during review, change evaluation, and problem analysis.

Tools and guidelines for meeting these challenges using technically sound processes are presented in parts II and III.

1.3.4 Phase IV Approve the Specification

A newly drafted specification must then be distributed to the parties who must abide by it. It is considered good practice for the preparer to highlight controversial issues so that they are not overlooked during the approval process. The reviewers are responsible for reading the document to ensure that it is correct, thorough, and clear. The producer has an additional responsibility to ensure that it has the capability to meet the requirements. This includes not only the product's performance, but the capability to test for any required property. The acceptance of customer specifications also requires approval. The process is less complex when an existing product meets the customer's specified requirements.

Sometimes it is useful when dealing with expensive test methods and materials to keep those who must accept the specification informed during specification development. This saves time because of the hierarchical nature of the processes. The testing costs should be determined before the target and specification limits are chosen in case excessive testing costs will prevent acceptance of the specification.

A specification is said to be approved when acceptance has been achieved. Conflict resolution may be required. It is important for all involved to know who has the ultimate authority for resolution.

1.3.5 Phase V Distribute and Maintain the Specification

Specifications are working documents. They must be available to those who need them, including the persons making the decisions to accept or reject product for release to a customer and those making the decisions to accept or reject upon receipt of a shipment. Specifications require periodic review. Document control ensures that any changes will be recorded.

CHAPTER 2

Types of Specifications

2.1 Classification of Specifications

Specifications can be categorized as internal or external and as process or product. Each of these four categories is discussed in this chapter. Both process and product specifications can be either internal or external. Thus,

	Process	Product
Internal		
External		

2.2 Internal Specifications

Internal specifications are generally restricted for internal use because they often contain proprietary information. Examples of internal specification requirements include manufacturing recipes, blending formulae, internal transfer methods, standard cost information, final product properties, and composition. Internal specifications are sometimes referred to as *one-party specifications* because they are developed and used within a single company.

Internal specifications often represent requirements within a company between internal producers and customers, where one group or department works to meet the needs of the next group or department in a manufacturing process. For example, the receiving department's customer is the production department. The receiving department's job is to have a sufficient supply of conforming raw materials to allow the production department to meet its schedule. The production department's customer is the sales department. In each case, the internal specification is a quantifiable expression of the (internal) customer's expectations and the (internal) producer's ability to meet those expectations.

2.3 External Specifications

External specifications are used between companies to document the requirements of a product or service. They are commonly used in supplier-producer and producer-customer relationships. Three categories of external specifications are

1. One-party specifications (developed by a producer) for an anticipated market
2. Two-party specifications (developed between two companies)
3. Third-party specifications (developed and controlled by a party outside of the supply chain, such as a professional society or a regulatory agency)

Examples of external specifications are sales specifications, customer specifications, raw material specifications, government regulations, Food Chemicals Codex, Food and Drug Administration (FDA) regulations, ISO standards, United States Pharmacopoeia (USP), or American Society for Testing and Materials (ASTM) testing requirements.

2.4 Product Specifications

Product specifications are tied directly to the finished product. They can be used internally for designing process-control strategies and for classifying product by grade. They can be used externally for specifying required properties of raw materials or useful properties of products for sale. Examples of product-specification content include physical or chemical properties, composition, performance, packaging, and delivery.

A customer may only have a product-performance requirement. The customer's concern is that the product must perform in a certain way when used in the customer's process. In that case, it is up to the producer to determine which finished-product properties need to be specified to ensure correct performance in the customer's process.

Product specifications should have as their basis the customer's requirements of the product. The producer's sales or marketing functions should obtain customer or market information on the performance profile of an approved product. The producer may then select finished product properties that will meet the required performance.

A specification is often identified as temporary until production samples of the proposed product have been evaluated and approved by the customer.

2.5 Process Specifications

Once appropriate product specifications are defined, a producer should establish process specifications to meet the required product characteristics. Process specifications are typically set points and ranges or instructions used to control a process in a way that will produce product that meets the customer's requirements. Examples of parameters that may need to be specified include temperature, pressure, material quantities, flow rates, and reaction times.

Process specifications are normally internal. A customer may, however, specify some process-control requirements, such as the sharing of control charts, sterilization, pasteurization, emulsification, and notification of significant process changes.

2.6 Other Product Specification Terminology

There are many names applied to various specifications within the CPI. Individual companies may define specific terminology to meet their own needs. It is important, however, that any specification be based on clear operational definitions and be internally consistent. The following terms are commonly used in the CPI.

- *Raw material specifications* (also known as purchase specifications) describe the requirements for incoming materials that will be used in the producer's process. In addition to product characteristic requirements, they may include specifications that describe requirements for product packaging and labeling.

- *Delivery specifications* describe requirements for the date, time, and place that the product is to be delivered. They may also include special handling and transfer requirements.

- *Sales specifications* are commonly used for commodity chemicals where producers define product characteristic ranges within which their product is sold. The intent is to define a product for a particular customer or market end use.

Producers often include typical values as part of the sales specification. (See sections 2.7 and 7.7.) It is not good practice to put typical values in a specification. Typical values belong in the technical data sheet. Typical values can be misleading because there is no standard method of developing and interpreting them. Statistical information, such as average values, C_{pk}, or P_{pk}, is much more meaningful.

- *Manufacturing specifications* are sometimes used by a producer as comprehensive internal product specifications. They may include properties not important to all customers or not included in other product specifications. Often they are made up of the most stringent characteristic requirements of a number of customers.

- *Release specifications* may refer to an internal translation of a customer's raw material specification into a form that is more useful to the producer, including the customer's special requirements. Release specifications can include requirements that the producer feels are important to the customer, though not formally cited in the customer's raw material specifications. Release specifications may include acceptance limits that have been computed to manage the risk of decisions made in the presence of variability. (See chapter 12.)

- *Temporary specifications* are also called *interim specifications* or *provisional specifications*. They are sometimes used during new product trials or when there is not enough data to predict long-term performance. New or development products are often initially sold on temporary specifications. Document and version control are very important for temporary specifications because they are subject to change.

2.7 Technical Data Sheets

Producers often prepare informational material containing typical values for product properties. These technical data sheets are not specifications. They do not include or guarantee ranges or specification limits. Technical data sheets are communication tools to provide information to potential users. Formal specifications need to be negotiated between the producer and the customer before the actual purchase of product. (See section 7.7.)

CHAPTER 3

Generalized Content of a Specification

3.1 General

Different kinds of information and requirements may be contained in specifications. They need to contain more information than only product properties and their numerical ranges. It is important to remember that specifications are communication tools. Thus, they need to provide information that is understandable by those who use them, such as the buyer, sales representative, laboratory technician, production operator, and raw material inspector. Several sample specifications are included in the case study in chapter 24.

3.2 Content

Specification format depends on the information required. No single specification will contain all the entries listed.

- General information
  - Company and/or organization name
  - Specification identifier
  - Revision number/date
- Approval signature or mechanism (section 23.2)
  - Producer's location (where appropriate)

- Process name
- Purpose and scope
- References to other documents

- Product description
  - Product name (trade name, grade)
  - Chemical Abstract Number
  - Contamination restrictions
  - Product unit definition
- Product characteristics
  - Characteristic name
  - Target and limits
  - Unit of measure
  - Test method
  - Significant figures
- Evidence of product conformance
  - Product certification
    - —Certificate of analysis
    - —Certificate of conformity
    - —Certificate of compliance
  - Product statistical information
    - —Control charts
    - —Product and process statistics
    - —Statistical quality control (SQC) information
  - Pre-shipment samples
- Control point identification
  - Target and limits
  - Unit of measure
  - Test method

  - Significant figures
  - Corrective action
- Sampling plan
  - Characteristic name
  - Sampling method
  - Frequency
  - Chain of custody
  - Resampling or retesting
- Boilerplate requirements
  - Producer to maintain effective quality system
  - Notice of change
  - Quality audits
  - Right to source surveillance
  - Records requirements
  - Deviations
  - Specification revisions
  - Confidentiality
- Packaging and shipping requirements
  - Transportation requirements
  - Packing list
  - Package marking or labeling
  - Bill of lading
  - Container material
  - Seals
  - Overpackaging
  - Previous content restrictions
- Regulatory requirements and standards
  - Quality system standards (ANSI/ISO/ASQC Q9000 series, American Petroleum Institute [API] Q1, and so on)

- Responsible Care™ (Chemical Manufacturers Association)
- Federal, military, FDA, Alcohol Tobacco and Firearms (ATF), and so on
- Other requirements, such as the QS-9000 automotive standard and Underwriters Laboratory
- Health, safety, and environmental concerns

CHAPTER 4

Specification Review and Approval Processes

4.1 Scope

The specification review and approval process includes both the adoption of new specifications and the revision of existing specifications. It applies to raw material, process, packaging, customer, product, label, or any other specification typical of the CPI.

The specification review and approval process is intended to serve each part of the supplier → producer → customer chain to ensure that the specification can be met, is properly defined, and that changes do not have an adverse effect on any part of the chain. The specification review and approval process must consider external requirements imposed by regulatory bodies, standard-setting organizations, and professional society requirements.

4.2 Review of Specifications

The review of a specification for an entirely new product or process requires comprehensive approval. This is often called a *commercialization procedure*, because there may be several approval stages involved. The review of a revised specification depends on the nature of the change. Typical changes are

- Adding or deleting a characteristic
- Changing a raw material or supplier

- Packaging and handling materials (bag, drum, and other discrete container requirements, pallet requirements)
- Label requirements
- Customer-marking requirements (special markings and labeling for specific customers, customer codes, and so on)
- Manufacturing process
- Information transfer (certificate of conformance, certificate of analysis, bill of lading, and so on)
- Test methods (revision of existing test methods, approval of new test methods)
- Logistics (shelf life, special transportation, shipping temperature, co-ship or pre-ship samples)
- Sampling plans
- Manufacturing site

As part of a commercialization procedure, specification documents may have temporary status while data are gathered and evaluated. This status allows for an abbreviated approval during commercialization to respond to incoming data. Such specifications should be clearly identified as temporary.

Because new or revised product characteristics may have far-reaching effects on the supplier → producer → customer chain, the approval process needs to be rigorous to ensure that the following issues are resolved.

- Is the process capable of meeting the requirements? (See chapter 15.)
- Do the requirements ensure fitness for use for the customer (internal or external)?
- Are raw materials available and capable of meeting the new or revised requirement(s)?
- Are regulatory or other third-party requirements met?
- Do all the participants in the chain understand the provisional nature of temporary specifications?

- Will there be any detrimental, unacceptable, or unexpected change for the customer?
- When and how will customer notification occur?

Resolution may involve balancing desired requirements such as the following:

- Testing: test accuracy and precision vs. response time
- Packaging: protection of product vs. package disposal and recycling
- Raw material: cost vs. grade
- Economics: cost vs. volume
- Logistics: frequency of production vs. storage stability

Therefore, the review and approval process for specifications for product-characteristic requirements needs to obtain agreement among the various functions potentially interested or knowledgeable of the issues.

4.3 Functions for the Review Process

Multiple functions provide important contributions in the review process.

• The *technical* function considers technical aspects of specification. This function evaluates whether a test is measuring appropriate variables; determines whether raw materials are appropriate and of appropriate quality; and applies knowledge and experience of similar processes, products, equipment, and measurements to current specification.

• The *production* function determines that the production process is capable of producing product that meets the requirements of the specification. This function evaluates the production economics and assesses response time and reliability of measuring instruments and test methods, which may or may not involve laboratory measurements.

• *Laboratory and testing* determine that testing capability exists to measure the process and/or product with the accuracy and precision required by the specification. These functions also determine that testing results can be available in the required time.

• *Production and/or test laboratory* determines that sampling capability exists to properly sample the process and product.

• *Packaging and handling* determine the capability to package and handle the product so that the requirements of the specification will be met when the material is received and used by the customer.

• *Marketing, sales, or customer* function determines that the specification meets the customer's needs as defined by marketing, sales, or the customer.

• *Quality assurance* determines that the quality system is capable of ensuring that the specification will be met.

• *Purchasing* determines the capability of the suppliers to meet requirements of the raw material specification.

• The *regulatory* function determines that regulatory or third-party requirements are not violated or compromised. Examples of regulatory agencies include the FDA and the Environmental Protection Agency. Third-party regulatory agencies include the Society of Automotive Engineers (SAE) and ANSI/ISO/ASQC Q9000.

4.4 Approval Authority

Approval is a commitment to meet the specification. It signifies that all the people who have interest in the manufacture, sale, and use of the product or process have considered their needs and potential compromises and have reached an agreement.

The management, or a designated representative, of all the functions whose work can either make, measure, preserve, or otherwise affect quality of the product covered by the specification should participate in the approval process. Establishing an owner of a specification can facilitate the approval and change process and serve to maintain the integrity of the information in the specification.

4.5 Change Management

The producer must clearly state the authority and functions required for approvals as part of the quality system, under *change management* or a similar heading. (See chapter 21.) There should be appropriate procedures and instructions on how specifications are changed and approved.

The change management procedure and instructions should be flexible enough to handle the approval needed for complex or simple specifications. These procedures and instructions should also cover temporary specifications. The provisional period and conditions of change in temporary status must be defined and understood by those affected by the specification.

4.6 Notification

Since there is risk of adverse or unexpected impact of any new or revised specification, provision must be made in the quality system for the notification of interested parties for new or revised specifications. Appropriate notification both inside and outside the organization must be defined. The notification should be commensurate with the nature of the change, the risk involved, and good quality practice. Appropriate lead time between notification and actual implementation of the specification change should be defined. The following groups may need to be notified.

- Customers
- Distributors
- Sales
- Marketing
- Production
- Laboratory
- Supplier
- Shipping
- Internal regulatory groups
- Technical groups

CHAPTER 5

Document Control

5.1 General

Document control involves having a system in place to ensure that written documents are properly developed, identified, maintained, approved, stored, distributed, and retired. Specifications and related documents should be under document control. If a contract so stipulates, then the requirements for document control can extend into the relationship between parties. ANSI/ISO/ASQC Q9000 requires document control of all quality system documents, including specifications.

Each specification should be clearly and uniquely identified to facilitate document control. As a minimum, the identification should include the following:

- A unique number or identification code
- A unique revision code
- Effective date
- Pagination

It is helpful if a "supersedes" date (effective date of the previous revision) as well as a "revise by" date are included.

5.2 The Supplier

When dealing with suppliers, the producer takes on the role of the customer. The producer's document control system should ensure that current specifications for relevant products are distributed to the supplier. The system should include a means to record that the supplier has accepted the current raw material specification. The producer's system should officially notify the supplier when the raw material will no longer be purchased. The supplier needs a system to retire the specification.

5.3 The Customer

When dealing with customers, the producer takes on the role of the supplier. The producer's document control system should ensure that either the customer has accepted the producer's sales specification or the customer's specification is current and accepted by the producer. The producer's system should officially notify the customer of the intention to no longer supply product to an existing specification. The customer needs a system to retire the specification.

5.4 The Producer

In addition to managing supplier and customer specifications, the producer's document control system should ensure that all specifications are current, properly approved, and distributed to the required locations within the company (for example, order entry and shipping control). When the producer's ordering system or shipping system is computerized, appropriate security control (for example, using passwords) and change documentation are needed to ensure proper document control and information transmittal. The effective date of a change may affect the shipment date as well as the manufacture of new inventory. All specifications and technical data sheets should be maintained under document control.

5.5 Record Keeping

All parties should keep complete historical records for each specification, from the initial negotiation stage until the product is no longer produced or purchased. The records should be kept for the time period required by internal, record-retention policies or regulatory requirements. These records include the following:

- Specification
- Approvals
- Revisions
- Acceptance notifications
- Negotiation
- Supporting data from testing and production

Part II

Development of the Specification

CHAPTER 6

Selection of Requirements

6.1 Introduction

In the CPI the most important characteristics address the processability and the end-use performance of the product. For chemicals the characteristics are often measures of the reactivity of the class of materials. Additional characteristics provide feedback to the process for control and improvement.

6.2 Reporting Characteristics

When possible, the characteristics should be specified with ranges and targets and should be expressed in standard units of measurement. For example,

- Sunscreen protection is measured by a sunscreen protection factor, a measure of the time of exposure to ultraviolet rays before skin redness results.
- The acidity is a measure of reactivity for carboxylic acid. This is commonly measured in acid value (mg KOH/g of sample).

Frequently these characteristics are measures of the purity of the product (for example, percent glycerol, amide content, or unsaturated fat content).

6.3 Product Specifications

Product specifications are usually confined to defining the product in standard units of measure for the producer's intended customer base. Product specifications may not contain information on characteristics specific to a customer's use. For example, the moisture content of bread is important to the shelf life of bread and desirability at the table. If the customer buys bread to make dressing for a turkey, moisture content may not be that important.

6.4 Raw Material Specifications

Raw material specifications define characteristics that affect the particular producer's processes. These specifications may include many characteristics and conditions besides those describing the product itself. Raw material specifications may be based on a supplier's sales specification, but may exclude characteristics unimportant to the producer's specific use.

6.5 Categories of Requirements

6.5.1 Physical Characteristics

Physical characteristics are commonly measured and specified. A frequently selected characteristic is some measure of color. The usual measurement is a comparison with a standard of reflected or transmitted color. Viscosity, specific gravity, boiling point, melting point, and cloud point are common characteristics for specifications in the CPI. The "mouth feel" of premium ice cream is a common physical characteristic measured by sensory testing. Odor, or lack of odor, is an important characteristic for cosmetics.

Contaminants and additives are specific to the manufacturing process and the proposed use. Moisture, heavy metals, and foreign matter are examples of undesirable contaminants in some products that may be selected for specification. Additives may include antioxidants, emulsifiers, flavors, vitamins, or perfumes.

6.5.2 Proprietary Characteristics

Proprietary characteristics or test methods may be very important to the product or service. These measures of performance should not be included in external specifications without secrecy agreements.

6.5.3 Packaging Requirements

Packaging requirements are important characteristics and need to be specified. For example, the delivery temperature of the contents of a tank truck is often included in the specification. Packaging requirements also include inert blanketing or antifungal fumigant requirements. Tamper-evident seals are important requirements for pharmaceuticals and kosher products. Special packaging requirements such as stainless steel containers, four-ply kraft paper bags, or polypropylene bulk bags may be specified. Minimum acceptable packaging, labeling, or other conditions should be specified. Some packaging requirements may be quite precise, defining static dissipation devices, heights, weights, volumes, or package materials. Labeling, coding, information content, and sealing requirements are frequently specified along with package requirements.

6.5.4 Handling Requirements

Handling requirements cover storage and delivery. Typical storage requirements are temperature and humidity restrictions, such as storage in an inert or nonreactive atmosphere or protection from solar heat or sunlight.

Delivery requirements may include special equipment to unload, a list of acceptable carriers, unloading by pressurizing with a nonreactive gas, seals on packages or trailers, pallet and strapping requirements, or the mode of delivery of shipping documents (for example, with driver). Any special equipment required for off-loading carriers must be specified.

6.5.5 Shelf Life Requirements

Shelf life declarations define the time over which a product is expected to meet specification. The shelf life is usually attached to recommendations for handling and storage conditions. Shelf life statements should be based on storage data for the material in containers similar to those being used and under expected or worst-case conditions.

6.5.6 Information Transfer

Requirements for information transfer may be included in a specification. A certificate of analysis or other medium of affirmation may be required to verify that the product meets specified requirements. (See section 22.3.) Information requirements may be specified as minimum values. These requirements frequently define the information to be included, the means of delivery of the information, and the timing of receipt. Information transfer requirements may include sample retention as well as record retention. This may include bar coding, the transmission medium for the certificate of analysis, or electronic data interchange requirements.

CHAPTER 7

Specification Values—Targets and Limits

7.1 General

Targets and limits are essential elements of a specification requirement. The target denotes the ideal value of the characteristic. The limits specify the permitted variation around the target. The target and limit values are a precise definition of the requirements and should be expressed numerically where possible.

7.2 Targets

The target for a characteristic should be set before the limits are set. The target for characteristics in a product specification should be set at a level that best meets the customer's need. The customer's need must reflect both technology and cost. The targets for internal characteristics in internal specifications should be aligned to produce the required final product. (See chapter 8.)

The measured values for a characteristic for all units of product will not be identical because of variability from lot-to-lot, sampling within a lot, and measurement. The Taguchi loss function (Figure 7.1) emphasizes that the farther the characteristic values move away from the target, the more cost is incurred by both the producer and the customer because of the additional complexity of compensating for the deviation. In order to

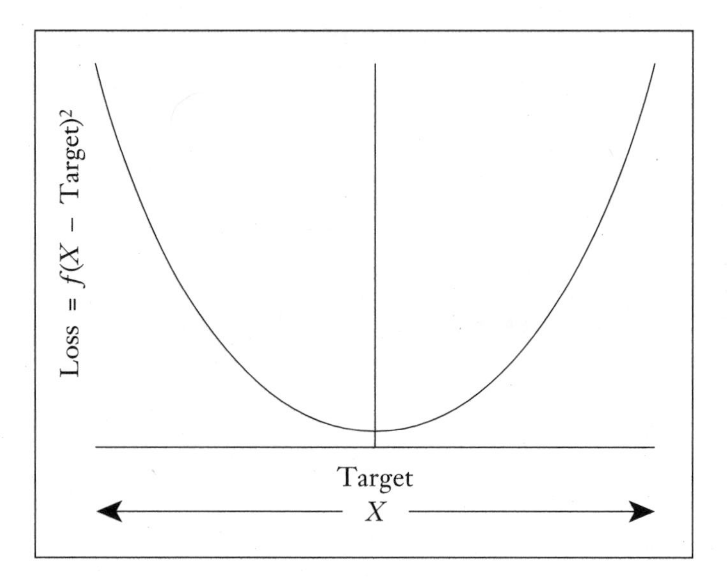

Figure 7.1. Taguchi loss function.

control the cost of the variability, specification limits are set about the target. (See Kackar 1985.)

7.3 Specification Limits

The term *tolerance limits* has been used to describe limits that define the conformance boundaries for an individual product unit. This meaning of tolerance limits is synonymous with the term *specification limits* used in this book. The authors have chosen to use specification limits because this terminology is more widely understood by nontechnical users. (See the Glossary.)

When customer needs have not been determined, or the product has performed satisfactorily, then the specification limits should be established so that nearly all of the distribution of characteristic values is included in the limits, regardless of the type of statistical distribution (for example, normal, lognormal, Weibull, and so on). The limits should be related to the target value of the characteristic, as in Figure 7.2.

Distributions are not always symmetrical. In Figure 7.3 the distribution of the characteristic is not symmetrical. The target has been located

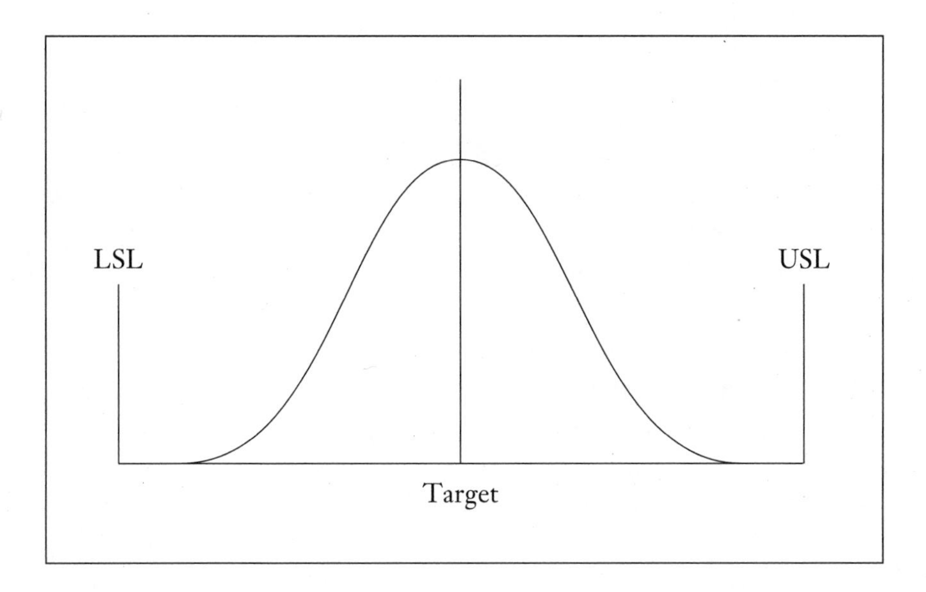

Figure 7.2. Two-sided specification limits—target centered.

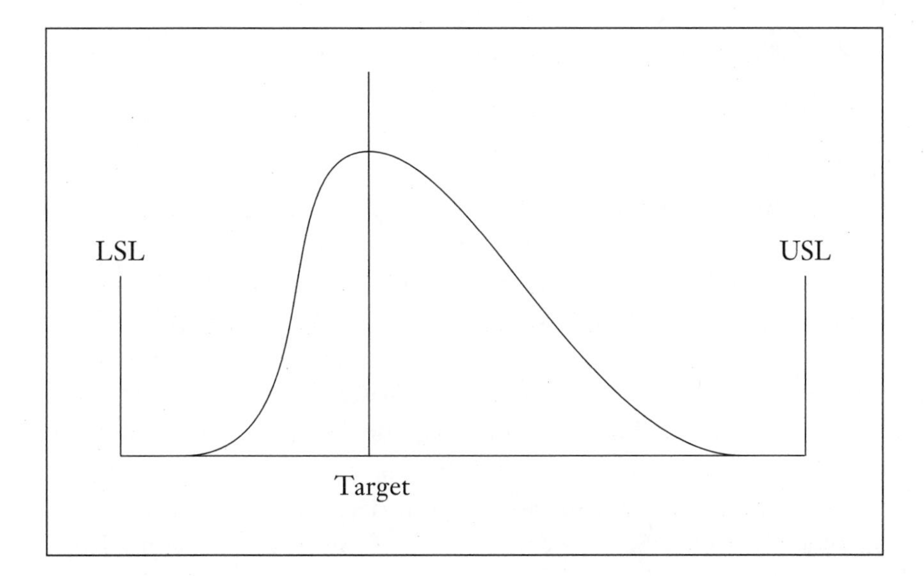

Figure 7.3. Two-sided specification limits—target not centered.

off-center to allow the distribution of the product characteristic to fit within the specification range. Nonlinear transformations can be applied to some nonsymmetrical distributions to improve the symmetry. When this is possible, the transformed variable should be used to determine the specification target. The logarithmic transformation and the square-root transformation are commonly used. Other transformations can be used; however, their use may require statistical expertise for valid application. (See chapter 20 of Natrella 1963.)

7.4 Statistical Limits

For a process that is in statistical control and for which the individual (or transformed) values follow the normal distribution, the location of the specification limits is often set as a multiple of the standard deviation of the individuals, s. (See section 10.2.) The specification limits should be set no less than ±4s from the target. (See chapter 14.)

Accepting specification limits closer to the target than ±4s may require either sorting of the product or a process change to reduce the process variability, in order to maintain process yield levels. These options incur additional cost.

7.5 One-Sided Specifications

Deviations from target are of concern only in one direction for some characteristics. Figure 7.4 shows a one-sided specification. One-sided specifications are bounded on the other (inactive) side by the laws of nature or measurement capability. Thus, they are a special case of two-sided specifications. One-sided specifications are common in the CPI. For example, assay or strength may have only a minimum limit specified, while impurities or defects may have only a maximum limit specified.

While a specification limit on one side of the target may indeed be adequate for the purposes of product release, two-sided specification limits are preferred wherever feasible because they focus the attention of the producer on the need to control variability.

In defining one-sided specifications the target value should be based on an analysis of data. The data may be historical or may come from an experimental design study. (See Box, Hunter, and Hunter 1978.) A

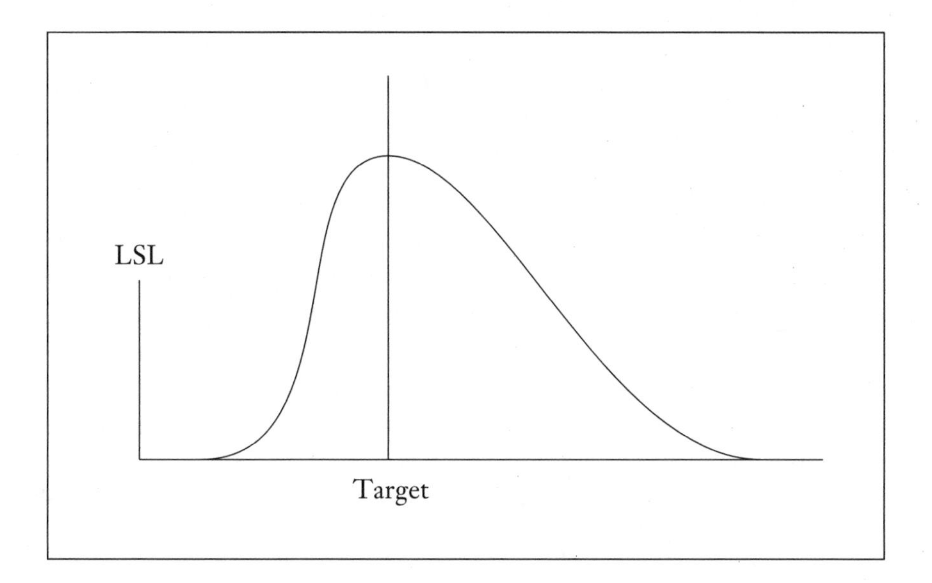

Figure 7.4. One-sided specification limit—with target.

properly designed, experimental design gives superior data and provides better information for setting specifications. However, acquisition of appropriate data and setting the limits is not a simple matter, and cookbook formulas are rarely appropriate for one-sided specification limits. The statistical distributions of characteristics, which may require one-sided specifications, are usually skew and may require statistical expertise for a valid analysis.

7.6 Pass/Fail Specifications

The pass/fail specifications are used for qualitative and semiquantitative determinations. For example, in order to pass, a sample must be "not cloudy at –10°C" or have "less than 10 ppm heavy metals."

Pass/fail specifications are often an economical way of supplying the information required by either the customer or the producer. For instance, it may not be worth the extra cost to determine the actual content of a heavy metal—the information that the material is "<10 ppm heavy metals"—may be sufficient. (See section 9.8.) Converting a numerical test result to a pass/fail value, however, discards potentially valuable information. Good practice calls for the use of numerical values if they can be obtained economically.

7.7 Typical Values

Typical values often appear on technical data sheets. These are not specification limits, but are the nominal values of selected characteristics. (See section 2.7.) There is no generally agreed standard for typical values. They may be expected minimum, maximum, or average values. If the customer has requirements for product characteristics, then they should be included in a specification, because typical values carry no guarantee.

Typical values are not necessarily tested on a routine basis and are not guaranteed on each shipment. It is implied, however, that the values of such characteristics for any shipment will usually be consistent with typical values.

7.8 Influencing Factors

Specification limits involve a balance between the customer's needs and the producer's process variability. Determining the robustness of the customer's process and the producer's process variability is critical to the development of good specification limits. (A process that is robust can tolerate more variability than a process that is not robust.)

Specification limits set before sufficient process performance information is available should be labeled temporary. Temporary specification limits should include a requirement to review the limits when data are available.

In addition to variability, the economic constraints of the processes of customer and producer are important criteria for setting specification limits and target values. Yields and environmental requirements are important economic considerations. Specification limits that are too wide may make the product uncompetitive. Specification limits that are too narrow may make it a specialty product. The process improvements to meet narrow limits may require capital investment by the producer.

CHAPTER 8

Specification Alignment

8.1 General

A product can be defined in many different ways depending on the perspective chosen. The sales representative may define it in terms of its application. The quality control technician may define it in terms of the final product test specification, while the production operator may define it in terms of process flows, pressures, and temperatures. To the outsider the defining characteristics selected by these individuals may appear to be describing completely different products. They are, however, all aspects of the same product.

8.2 Hierarchical Specifications

As pointed out in chapter 2, many types of specifications are used to define requirements of a particular product. Although the types serve different purposes, they are dependent on each other and have a hierarchical relationship. If each type of specification represents a plan to produce an acceptable final product, then the makers of each specification must take into account all of the design needs for the final product.

Recognizing the supply chain concept, the hierarchy proceeds backwards from the customer to the producer to the supplier as follows:

1. The targets for the processing and end-use performance requirements are designed to meet customer and market needs.

2. The producer's product specification must define targets for appropriate measurable product characteristics in order to achieve the targets for customer processing and end-use performance characteristics.
3. The manufacturing and raw material specifications must define targets and appropriate measurable production conditions in order to manufacture a product aimed at the targets of the product specification.

It is not uncommon for many of the characteristics to be the same from one level of the chain to another. In this, some specification requirements can be used without additional development effort.

8.3 Alignment

The targets in each specification must be aligned from one type of specification to the next. In addition, the limits in each specification must take into account both the variability and shift in level of the process characteristic. If this alignment is successful, then the product design process culminates in a product that can be manufactured successfully, characterized fully, functions in the customer's manufacturing process, and satisfies end users. On the other hand, the failure to align the targets of the characteristics in the product design process exposes the producer to increased costs for nonconforming product, downgrades, and complaints.

Figure 8.1 illustrates the relationship between product development and specification management. The purpose of the figure is to provide an overall picture of the process. Some steps may be detailed and may involve a number of quality technologies. The flow proceeds from the upper left corner, down the page, across the bottom, and up the right side. On the left the four development steps proceed from the customer's need to the manufacturing plan. At each stage the requirements are redefined in terms of the process involved. Note that the type of the specification is indicated in the oval below and to the right.

The verification processes are indicated on the right. At each of the three verification steps the conformance of the product is compared to the requirements. If the product misses requirements, then it is returned to the corresponding development step. When the product meets all the requirements, the specifications have been aligned and the product is ready for commercialization.

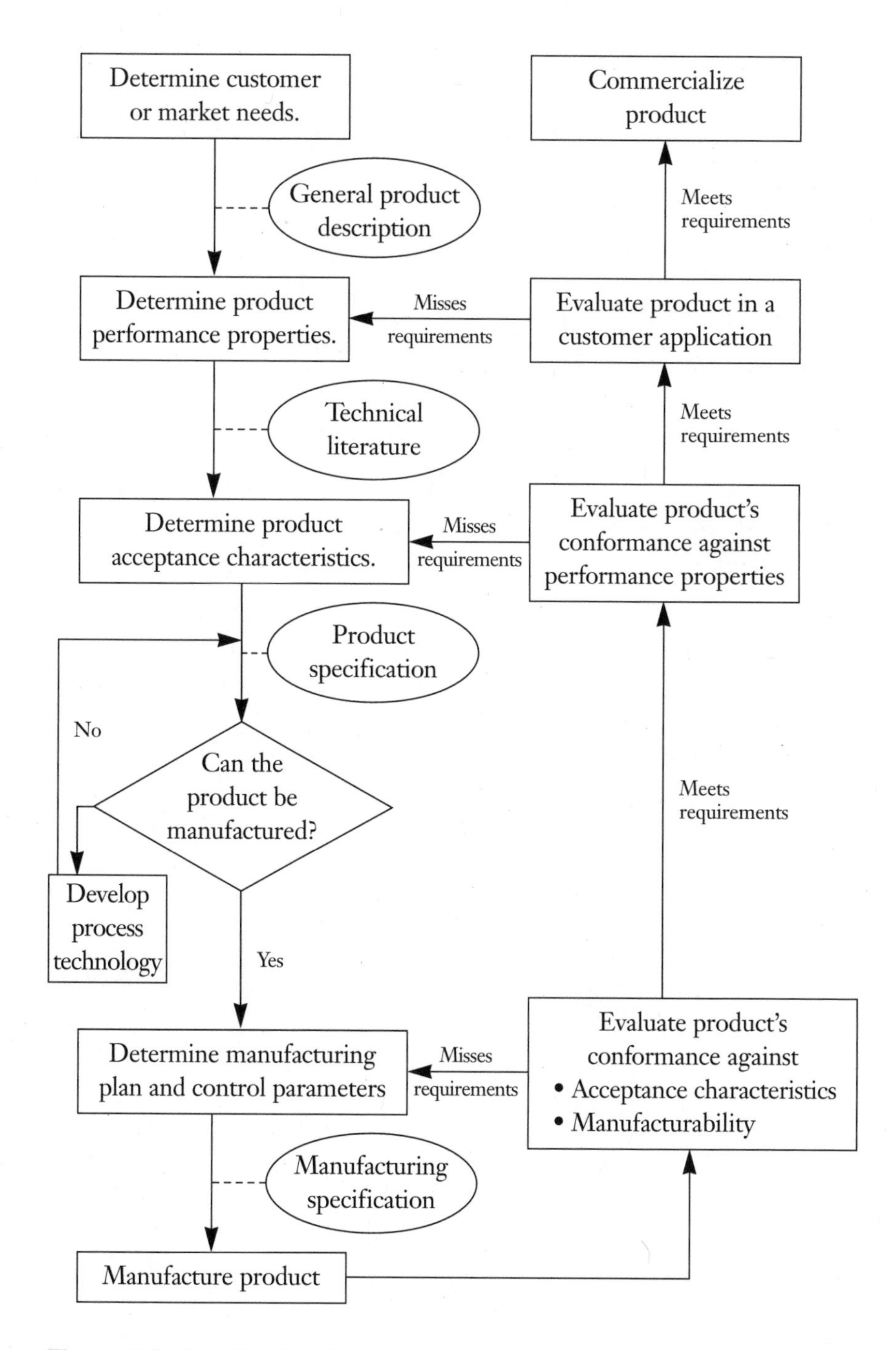

Figure 8.1. Specification alignment in the product development process.

CHAPTER 9

Test Methods and Measurement Processes

9.1 Issues

A specification should identify the test method used to measure the characteristic or process parameter. In the CPI the inherent variability of the test method is usually the largest contributor to the measurement process variability, although other sources also contribute. (See section 10.3.2.) The variability of the measurement process must be known in order to arrive at correct decisions with respect to specifications. As key contributors, all measurement processes must be managed as part of the quality system.

9.2 Standard and Reference Methods

Many industry groups have standard test methods. Where they are capable, they are preferred because they provide a consistent basis for comparison of product quality over time and among different producers. If the test method is used exactly as documented in the standard, the requirement for test method sharing is usually covered by the standard.

Standard test methods are often modified by a company because of the equipment available or because it is necessary to make the method more specific in practice to achieve the test variability required. For example, a standard test method might use nonspecific terms such as *acidic* or *basic*. Specific terms, however, may be needed, such as pH = 8 ± 1, to conform to good laboratory practice. A modified, standard test method requires the same validation and measurement assurance information as any internal test method.

If a standard test method has been modified, it is no longer the standard test method, and the standard test method becomes the reference for the method actually used. The actual test method must be indicated in the specification, although the standard test method can be referenced to document the original source of the test method. (Example: Test Method ASTM xxxx at time = 25 min., pH = 8.5.

9.3 Internal Methods

Producers are excellent sources of test methods. They have usually performed the tests frequently enough to provide sufficient data for the determination of measurement process capability. (See Nelson 1995, or Wheeler and Lyday 1989.) Producers should be willing to share test methods and test capability data with customers.

When internal methods are used, they must be supported by measurement assurance procedures. The originator of an internal method has the responsibility to provide capability data. Including measurement-process capability in a specification may be helpful to communicate the amount of measurement variability contained in the specification limits. (See section 15.4.)

9.4 Performance Tests

Performance tests are scaled versions of an actual production process application. Although they may be of low precision, they may be the best indicators of suitability for use. The requirements for test method capability also apply to performance tests.

9.5 Measurement Assurance Methods

Test methods should be evaluated for range, accuracy, precision, bias, ruggedness, and durability. The test method selected should ensure consistent decisions based on the test results. All of the characteristics of a test method need to be taken into account when using measurement data to set specification values and accepting product.

Each measurement process needs to operate in statistical control, which may be demonstrated by use of check samples, control charts, or other statistical techniques appropriate to the test method used. Analyses, measurements, and tests should be performed in a manner that will provide results that are repeatable and reproducible to an extent consistent with the technology involved.

Where practical, traceability to reference standards of known value and stability, preferably to national or international standards, is good practice. In industries where such traceability does not exist, specially developed reference materials must be used.

Where practical, the customer and the producer should periodically participate in an established interlaboratory test program designed to evaluate the status of testing proficiency.

9.6 Alternative Test Methods

Some product characteristics can be measured by more than one test method. The accuracy and cost of alternative measurement methods should be examined in selecting the method for use. Alternative test methods may report results for the same characteristics in different units of measure. The relationship between alternative measures must be known and documented.

For example: The metal content of a material can be measured by inductively coupled plasma (ICP) spectrophotometry, atomic absorption (AA) spectrophotometry, or colorimetric test methods.

- ICP and AA are test methods that employ computerized measurement of atomic emissions for the quantitative detection of small amounts of each metal. The units of measure are μg/g.
- Most colorimetric tests are semiquantitative and only inform that the metal content is greater or less than a reference concentration. The test results are reported as "greater or less than the reference."

The appropriate test method depends on the decision to be made.

9.7 Units of Measure

Use of International System (SI) units of measure is recommended where practical. For some tests it is often desirable to use units common to commerce or regulatory bodies. (For example: Flash points are generally reported in °F.)

9.8 Detection and Quantitation Limits

The limit of detection (LOD) and limit of quantitation (LOQ) of a test method must be considered for low concentrations of impurities or additives. The following guidelines are taken from Taylor 1987.

The LOD is the smallest quantity of a characteristic that can be determined to be statistically different from the absence of the characteristic. The uncertainty of a measurement at the LOD is ±100 percent. The LOD can be estimated as $3s_0$, where s_0 is the standard deviation of a blank sample containing none of the characteristic. If the LOD is 3 ppm then measured result of 2 ppm should be reported as "<3 ppm."

The LOQ is the quantity of a characteristic that can be measured and is quantitatively meaningful. The uncertainty of a measurement is ±20 percent and increases rapidly approaching the LOD. The LOQ can be estimated as $10s_0$. Measurements below the LOQ should be regarded as qualitative. If the LOQ is 10 ppm, then a calculated measurement of 6 ppm should be reported as either "detected" or "3 to 10 ppm."

Both the producer and customer must agree on the LOD and the LOQ so that misunderstandings do not occur. Statistical expertise may be needed for an adequate evaluation of the sensitivity of a test method.

Specification limits should be set above the LOQ. If a lower limit is desired, then the measurement procedure must be improved to reduce the variability, or multiple measurements must be made to reduce the variability of the calculated average.

Part III

Data Collection, Evaluation, and Use

CHAPTER 10

Variability Issues

10.1 Scope

Variability is a major issue in the CPI. Its effect on quality decisions is often not understood. This chapter gives principles and guidelines for the CPI. It discusses the sources of variability and how they link to the various functions associated with specification values.

Understanding the effect of variability is important for many sections of this book, particularly the following:

- Chapter 11—Sampling Plans
- Chapter 12—Decisions Based on Data
- Chapter 13—Statistics for Decisions Based on Data
- Chapter 15—Capability and Performance Measures
- Chapter 16—Conformance Decisions
- Chapter 24—A Case Study

There is seldom only one approach or rule that is best; however, by following a set of principles and guidelines that addresses the entire specification system, the user will have confidence that the effect of variability has been minimized.

10.2 The Standard Deviation

The basic measure of variability is the population standard deviation, σ (sigma). In practice, all the members of a population cannot be examined. Therefore, σ is estimated by sampling the population and measuring the variability of the members of the sample. This variability, called the sample standard deviation s, is defined as

$$s = \sqrt{\Sigma(x_i - \bar{x})^2 / (n - 1)}$$

where x_i is the measurement of *i*th member of the sample
$\bar{x}$ is the sample average
n is the number of members in the sample

The estimate of the sample standard deviation will be adequate for practical purposes, if n is equal to or greater than 30 observations. If action must be taken based on values computed from smaller samples, statistical expertise is needed to evaluate the increased risk involved.

Note: The word *sample* has two meanings in the CPI. Care must be taken to differentiate between the meanings in the context of the discussion.

- To the statistician a sample is a subset of members of a population, which is used to estimate by inference characteristics of the population.
- To the analytical chemist a sample is a physical amount of material selected for analysis.

Sampling will be discussed in more detail in chapter 11.

10.3 Why Variability Is an Issue in the CPI

The standard deviation is the key statistic for many calculations in statistical quality control. It is used to describe variability from a single source; however, variability in the CPI often comes from multiple sources.

The square of the standard deviation is called the variance.

$$\text{Variance} = \sigma^2$$

An important property of the variance is that variances are additive for different sources acting together in a system. This property permits the splitting of the variance into separate sources that are called *variance components*. The magnitude of these various sources can be discovered by analysis of variance (ANOVA). The two major sources are product variability from the manufacturing process and measurement system variability. These sources can, in turn, be further partitioned into subsources that have practical implications (Herman 1989, 670–675).

10.3.1 Product Variability

Product variability represents real differences in product characteristics that may be detectable by the customer. It can be split into two components—lot-to-lot variability and within-lot variability.

1. *Lot-to-lot variability over the long term*—Many CPI processes exhibit an inherent variability that extends over long production times. Some of the factors that can contribute to the inherent lot-to-lot variability in excess of the within-lot variability include the following:

- Raw material variability
- Transport, storage, and handling of raw materials
- Environmental—(ambient conditions, cooling water temperature, and so on)
- Long-term variability in continuous processes—(process aging, and so on)
- Semi-continuous process steps—(large batch blending, and so on)
- Equipment differences
- Personnel differences
- Batch-to-batch variability

If lot-to-lot variability is very large, product may have to be sorted to meet the customer's requirements, or the size of the lot may have to be reduced to cover a more homogeneous production period. Lot-to-lot variability may be reduced by the use of statistical process control (SPC) or the introduction of process improvements.

2. *Within-lot variability over the short term*—This component represents variability among units within lots. It directly affects sampling among units within lots. (See chapter 11.) If this component is large,

more physical samples will be required to obtain adequate precision for lot averages. Within-lot variability may be reduced by improved mixing or the use of automatic process control. The reduction of within-lot variability is often difficult and may involve significant technical effort.

10.3.2 Measurement System Variability

Measurement system variability includes the variability in the reported results for the entire measurement process from sampling through testing. Measurement system variability does not directly affect the performance of the product in the customer's process. Indirectly, measurement variability can affect the product by changing the performance of the SPC system in place. It does affect the assessment of product characteristics, however, because it is additive to the product variability. The variability of the test method itself plays a major part, but other factors also add to the measurement variability. Some factors that can affect short-term and long-term measurement variability are

Short-term

- Test method
- Sampling procedure
- Sample preparation
- Calibration (each sample)
- Within sample variability

Long-term

- Calibration (periodic)
- Multiple test equipment units
- Operators (sampling)
- Operators (testing)
- Ambient conditions

If the short-term measurement variability is large, multiple analyses will be required to obtain precise results. In order to reduce it, a detailed study of the measurement process may be required or a different test method used.

If the long-term measurement variability is large, there may be apparent disagreement among samples run on different days. It may be reduced by use of SPC on the measurement process, recalibration, or operator retraining. It may also be appropriate to examine the test procedure instructions for inconsistencies or ambiguities.

10.4 An Example for Obtaining Estimates of Variability

This section discusses a sampling, testing, and analysis plan for obtaining estimates of variability. There are many other plans that would work as well. (See Duncan 1986 and Schilling 1982.) A practical example using this plan is included in chapter 24.

The intent of this example is to convey the principles and concepts that should be followed. It uses a balanced experimental plan so that the estimates of variability are easily obtained. There are satisfactory smaller plans that are unbalanced. The analysis of unbalanced plans requires statistical expertise (Nelson 1995).

The following estimates of the components of variance described in section 10.3 are needed.

- Product variability
  - Lot-to-lot variability
  - Within-lot variability
- Measurement variability
  - Long-term variability
  - Short-term variability

The sampling diagram is shown in Figure 10.1. There are four levels of sampling. The top level represents the N lots. The second level represents the two distinct physical samples from the lot. The third level represents the two distinct test times. The fourth level represents the two replicate analyses run on each sample at each test time. Table 10.1 is the ANOVA table for this plan.

For this example, assume that lots have been rationally determined, and there are multiple lots within a given day. If the manufacturing process is continuous or close to continuous, then the best lot structure may not be immediately obvious, and a trial-and-error procedure may be required to find an appropriate one. Collecting data on a trial lot size may help. For example, choose a period such as eight hours for a lot. Take within-lot samples four hours apart. If within-lot variability is large relative to lot-to-lot variability, try a shorter period for the lot.

- *Lot-to-lot variability*—Enough lots must be sampled to provide a valid estimate of the lot-to-lot variability with reasonable precision. A sample size of 60 or more is preferred. A practical minimum is 30 lots. It is important that the time frame selected be chosen to

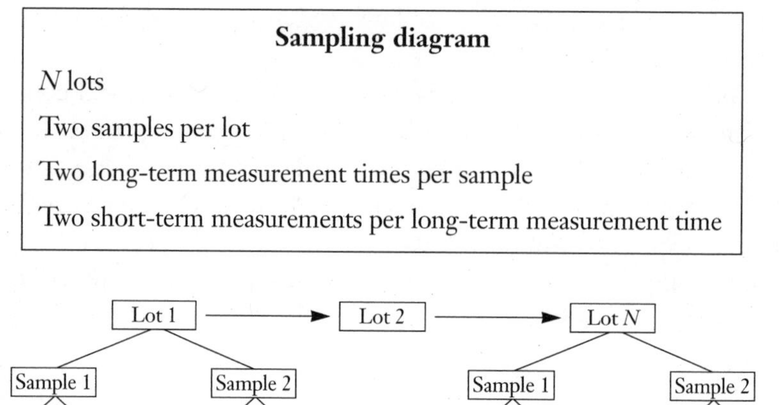

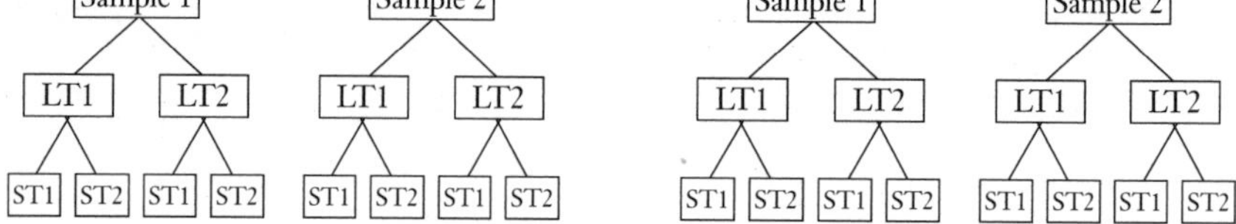

Figure 10.1. Sampling diagram.

Table 10.1. ANOVA for the sampling diagram.

Source	Degrees of freedom	Expected mean squares
Lot-to-lot	$(L - 1)$	$\sigma^2_{ST} + r(\sigma^2_{LT}) + lr(\sigma^2_{WL}) + lrS(\sigma^2_{LOT})$
Samples within lot	$(S - 1)L$	$\sigma^2_{ST} + r(\sigma^2_{LT}) + lr(\sigma^2_{WL})$
Long-term measure	$(l - 1)SL$	$\sigma^2_{ST} + r(\sigma^2_{LT})$
Short-term measure	$(r - 1)lSL$	σ^2_{ST}

where

- L = number of sampled lots
- S = number of samples per lot
- l = number of long-term measurement times per sample
- r = number of repeat measurements per long-term test time
- σ^2_{LOT} = Variance lot-to-lot (product only)
- σ^2_{WL} = Variance within lot (product only)
- σ^2_{LT} = Variance long-term measurement
- σ^2_{ST} = Variance short-term measurement

include most of the common sources of lot-to-lot variability such as start-ups, daily ambients, shifts, machines, operators, trends, cycles, and so on. A minimum production time of 90 days is suggested. For example, sampling a lot every other day for 90 days would satisfy the minimum requirement.

- *Within-lot variability*—At least two samples should be taken from each lot. A sample size of two was chosen for this example. If the lot is not time- or quantity-structured, the samples should be randomly chosen throughout the lot. If there is a time or quantity structure to the lot, then good information can be obtained from samples that are evenly spaced within the lot.

There are two components of measurement variability. In order to estimate them, take enough product to make four measurements from each within-lot sample. Test two of the measurement samples as soon as possible after sampling. Save the remaining two for testing at a later time.

- *Long-term measurement variability*—It has been assumed for this example that an adequate estimate of long-term measurement variability will be obtained by testing both samples labeled ST1 on one day and both samples labeled ST2 on the next day. In other situations, testing the ST2 on another shift, with a different operator, or on another piece of equipment may be adequate to obtain the desired estimate of long-term measurement variability.

- *Short-term measurement variability*—In order to obtain an adequate estimate of short-term measurement variability, the two samples should be tested under the same conditions (shift, operator, test equipment). They should, however, be treated as distinctly different samples including the sample preparation steps.

10.5 Estimating the Variance Components

The variance components are estimated by ANOVA. Nelson (1995) explains the procedure briefly. Hicks (1956) wrote a series of three articles that give a comprehensive discussion of the different ANOVA models, their variance components, and the computations required. In the ANOVA model each observation Y_{ijkn} includes the true value μ, plus adjustments for the lot LOT_i, for the sample within the lot $SAM_{j(i)}$, for the long-term measurement within the sample $LT_{k(ij)}$, and for the short-term measurement within the long-term measurement $ST_{n(ijk)}$. The model can be expressed as follows:

$$Y_{ijkn} = \mu + LOT_i + SAM_{j(i)} + LT_{k(ij)} + ST_{n(ijk)}$$

A general variance model has been defined to include the sources of variability previously specified. It shows that the total variance, V(Y), is

$$V(Y) = \sigma^2_{TOT} = \sigma^2_{LOT} + \sigma^2_{WL} + \sigma^2_{LT} + \sigma^2_{ST}$$

10.6 Understanding the Variability

The variability of an average that is subject to multiple sources of variability can be estimated by use of variance components. The variance of an average σ^2_{avg} can be computed as the sum of the ratios of the variance components σ^2_i, divided by the number of sampling elements for the source represented by the variance component n_i. In terms of the variance components defined, it is

$$\sigma^2_{avg} = \sigma^2_{LOT}/n_L + \sigma^2_{WL}/n_W + \sigma^2_{LT}/n_l + \sigma^2_{ST}/n_S$$

where

n_L = number of sampled lots
n_W = number of samples per lot
n_l = number of long-term measurement times per sample
n_S = number of repeat measurements per long-term test time

The standard deviation of the average is the positive square root of the variance. The importance of this formula is that it quantifies the intuitive understanding that averages have less variability than individual observations. In addition, it allows multiple sources of variation to be taken into account.

These expressions have been given in terms of the population components indicated by the Greek letter σ. For practical calculations the sample estimates indicated by the Roman letter s are used in place of the population components.

10.7 Useful Measures from Variance Components

Four useful measures of variability can be obtained from the variance components. They are

Grand total variance:	$\sigma^2_{TOT} = \sigma^2_{LOT} + \sigma^2_{WL} + \sigma^2_{LT} + \sigma^2_{ST}$
Total product variance:	$\sigma^2_{PROD} = \sigma^2_{LOT} + \sigma^2_{WL}$
Within-lot sampling variance:	$\sigma^2_{WLS} = \sigma^2_{WL} + \sigma^2_{ST}$
Total measurement variance:	$\sigma^2_{MEAS} = \sigma^2_{LT} + \sigma^2_{ST}$

The standard deviation for each of these estimates is the positive square root of the variance. While it is good practice to develop specification limits based on the product variability only, either the grand total or the total product standard deviations can be used (ASQC Chemical and Process Industries Division 1987). The grand total and the within-lot sampling standard deviations can be used to calculate performance and capability indexes. (See chapter 15.) The within-lot sampling standard deviation applies to most lot-sampling situations. (See chapter 11.) This estimate assumes that all measurements are to be made at the same time. If the measurements are to be made over an extended time, then the σ^2_{LT} component must be included.

The variance components can be combined in many ways to reflect the variability of averages and differences that may be computed in the development and the use of specification limits.

CHAPTER 11

Sampling Plans

11.1 Scope

Sampling plans must be used in the CPI. The prime objective of sampling is to obtain an estimate of a characteristic of a population so that a decision can be made regarding that population. Decisions are often required for product acceptance, product release, or process control in relationship to specification limits. The values reported from samples may be used to identify a product, or to estimate the level of potency of a component or the concentration of an impurity in a lot.

According to *Quality Assurance for the Chemical and Process Industries: A Manual of Good Practices* (1987, 13),

> Sampling is the process of removing a portion of a material for analysis. Ideally, the sample should truly represent the material from which it was taken. It must maintain its integrity with respect to the characteristics measured, and the act of sampling should not change the chemical sampled. Sampling should provide low cost measurement data on the material with known and controlled risks.

11.2 Population

The population being sampled is the defined quantity of material for which the mean level of some characteristic is being estimated. In the CPI this quantity is often referred to as the *lot.* A rational lot should be accumulated under process conditions considered uniform. Each batch is usually considered to be a lot for a batch process. A period of uniform process conditions during a specified period of time is usually considered a lot for continuous operations.

11.3 Estimation

Random samples taken from the lot may be evaluated for physical characteristics or subjected to chemical testing to estimate the mean level of a property or characteristic. The value of such estimates is dependent on the following:

- The amount of variability of the characteristic within the lot
- The measurement variability
- The physical sampling method

The magnitude of this variability can be estimated by variance component analysis. The within-lot sampling variance for single analyses on each of several test samples from a lot is

$$\sigma^2_{WLS} = \sigma^2_{WL} + \sigma^2_{ST}$$

See section 10.7 for more details. Alternatively, the standard deviation may be computed directly by pooling estimates from multiple samples within a number of lots. Simple repeated sampling within the lot does not permit the separate estimation of the measurement components of variance. The precision of the estimated mean can be improved by increasing the number of physical samples used to compute the mean.

11.3.1 Sample Size

A common question is, How many items or measurements should be included in a sample? The answer is dependent on the intended use of the results. In general, the question arises because of the desire to obtain a specified precision about an estimated mean with a stated degree of confidence. The degree of precision and the degree of confidence required may be part of a product specification.

A sample of one is often taken to characterize a lot in the CPI for economic reasons regardless of the precision. A sample of one may be adequate for qualitative testing for chemical identity; however, it may not provide adequate precision for release decisions. (See chapter 13.)

11.3.2 Confidence Limits

The confidence interval can be used to estimate the reliability of a mean $\bar{x}$. A confidence interval is an interval that will include the true population mean with a given level of confidence expressed as a percentage. A confidence interval can be computed using the Student's t distribution (Guttman, Wilks, and Hunter 1965).

$$CI_{1-\alpha} = \bar{x} \pm t s_{avg}$$

where t is the value of student's t for the $(1 - \alpha)$ level of confidence and k degrees of freedom; for a sample of size n from a lot $k = n - 1$. If the standard deviation is determined from other data, then k will be determined by the size and structure of those data.

s_{avg} is the standard deviation of the average.

Consider a numerical example that illustrates the effect of different sample sizes for average and for the standard deviation. A sample of size five has been taken from a rail car of polyethylene to estimate the melt index of the material. The reported values are 6.5, 7.4, 7.3, 5.3, and 5.9. The computed value of $\bar{x}$ is 6.48. The computed value of the sample standard deviation is 0.90. The computed value of the standard deviation of the average is $0.90/\sqrt{5} = 0.40$.

If the sample standard deviation is used, then the value of t is 2.78 for four degrees of freedom at a confidence level of 95 percent. The 95 percent confidence interval is

$$CI_{0.95} = 6.48 \pm (2.78)(0.40) = 6.5 \pm 1.1$$

If a long-term estimate of the sample standard deviation 1.10 from a laboratory control chart with 91 observations is used, then the value of t is 1.99 for 90 degrees of freedom at a confidence level of 95 percent.

The computed value of the standard deviation of the average is now $1.10/\sqrt{5} = 0.49$. The 95 percent confidence interval is

$$CI_{0.95} = 6.48 \pm (1.99)(0.49) = 6.5 \pm 1.0$$

11.3.3 Required Sample Size

The confidence-interval formula can be used in reverse to estimate a required sample size using the formula $n = (ts/D)^2$, where D is the desired difference from the mean. The value of D must be determined by the business needs of the problem. The basic question is, How large a difference from the mean must be detected for practical work? The calculated sample size is the number of samples that must be obtained when each sample is measured once. Rough estimates can be made using $t \cong 2$ for 95 percent confidence limits or for the sample size to give an approximate 95 percent confidence interval for the population mean. Consider the example above. If $s = 0.9$ and the desired D is 0.5, then

$$n = \left(\frac{ts}{D}\right)^2 = \left(\frac{2 \times 0.9}{0.5}\right)^2 = 12.96$$

Since the sample size must be an integer, the required sample size is 13.

The assumption for using the Student's t distribution for confidence intervals is that the individual measurements follow the normal distribution. While this assumption is valid in most circumstances, statistical expertise is required if the data are not approximately normally distributed.

11.4 Physical Sampling Considerations

It is important to use knowledge of the process, the chemicals involved, and their physical states when developing sample procedures. In particular, the physical state of the material samples is a vital issue. Solids, liquids, gases, and mixed-phase materials require different physical sampling strategies. Where possible, each sample should be analyzed in its entirety. If a smaller sample is required for the analysis, care must be taken to ensure that it is representative of the total sample (Juran 1974, Sec. 25-A).

11.4.1 Solids

- *Discrete units*—When in discrete units such as pellets, briquettes, bales, bags, rolls, or bundles, the representative sample should consist of units distributed throughout the batch so that within-lot variability is included in the sample. It is also essential to establish the degree of variability within large discrete units—for example, the bale—as well as between discrete units—that is, bales.

- *Bulk solids*—Bulk material may be natural (such as ores) or have been processed from an industrial process (such as powders).
    - *Natural solids*—Physical samples must be taken throughout the lot, pulverized and analyzed. Subsamples can then be taken to estimate the level of a characteristic and measurement variability.
    - *Processed solids*—For lot acceptance, if no information is available from the processing lot, samples should be taken throughout the lot, and tested to estimate level and measurement variability.

11.4.2 Liquids

Liquid samples from unmixed tanks should be taken at several different levels to detect any stratification. Circulating liquid in a tank does not necessarily ensure adequate mixing, and periodic samples should be taken. Mixing or stirring is commonly used for liquids.

11.4.3 Gases

In general, gases are considered to be homogeneous. Gaseous mixtures may become stratified if pressure and temperature are not properly controlled. Temperature must be maintained above the dew point of all components, and the critical pressure of no component should be exceeded.

11.4.4 Mixed-Phase Materials

Materials such as emulsions, gels, slurries, and so on are mixtures that create difficult sampling problems. Samples should be taken from several locations in the bulk container and analyzed. Care must be taken to avoid demulsification, separation, or settling in the sample.

11.4.5 Compositing

The compositing of multiple physical samples for analysis is commonly used to reduce testing cost. The compositing of physical samples is appropriate if

- The measurement variability is small.
- The within-lot variability is large.
- Only an estimate of the average is required.

Compositing is *not* good practice. It is appropriate only if all three conditions are met. Better practice calls for individual analyses on each sample. If the additional testing cost or volume is not acceptable, then alternative sample structures or test methods should be investigated.

An additional problem with compositing is that it conceals the within-lot variability. The absence of an estimate of within-lot variability may defer needed corrective action to reduce that variability.

11.5 Sampling for Process Monitoring or Control

Periodic sampling of a process stream prior to intermediate bulk storage or the ultimate container is commonly used for process control and adjustment. Such sampling may involve autosamplers or manual techniques. Process streams of solids, liquids, gases, or mixtures may be sampled in this manner. The data from periodic sampling can provide information for making lot acceptance or release decisions as well as for process adjustment. In order to use such data for lot acceptance or release decisions, the automatic sensor must be validated as accurate and reliable. It is good practice and an economical use of resources to base lot acceptance on such process-control data.

Where process control procedures involve periodic sampling of operating conditions but not of analysis, the periodic samples can be composited over the period of time defined as a lot. Acceptance can then be based on analysis of the composite, while noting the cautions about compositing.

11.6 Published Plans for Acceptance Sampling

Published sampling plans for attribute sampling, such as ANSI/ASQC Z1.4-1993 and Schilling (1982), can provide acceptable plans for inspection purposes. Attribute sampling involves evaluating the presence or absence of some characteristics or attribute in each of the units in a lot, and then counting how many units do or do not possess the quality attribute.

The most widely used procedure in acceptance sampling is the single sampling plan. Such plans are thoroughly discussed in the cited references. In addition, these references contain information on other useful plans such as double and multiple sampling plans. Such plans allow a lot to have additional chances to pass acceptance limits. These plans allow for additional samples to be taken to achieve more discrimination in the disposition of a lot. (See chapter 13.) For high-quality products in continuous or frequent production, these plans can be unduly expensive. An alternative is to consider skip-lot or isolated-lot sampling plans (ANSI/ASQC S1-1987 and ANSI/ASQC Q3-1988). The use of skip-lot sampling plans is good practice if lot quality is high. The complexity of changing sampling frequency when lot quality is variable is undesirable.

In the CPI, variable sampling, rather than attribute sampling, is more common. The standard variables sampling plans are ANSI/ASQC Z1.9-1993. These plans are similarly unduly expensive to use when high-cost material is involved. As quality systems, such as those in the ANSI/ISO/ASQC Q9000 family and automatic process control, become more widely implemented in the CPI, process capability improves, and these traditional sampling plans often become too expensive. Batch production becomes more uniform giving higher credibility to lot averages than for processes without quality systems. Continuous processes under process control also reduce the lot-to-lot variability. Thus, skip-lot sampling may be considered for some characteristics. The comments about skip-lot sampling for attributes also apply to variables sampling.

CHAPTER 12

Decisions Based on Data

12.1 Introduction

Measurement variability is present in data used for quality decisions in the CPI. Many decision makers do not understand the effects of measurement variability on the decisions they make. They are often willing to accept a single number (or average) as an adequately precise estimate of the true value of a measured characteristic. It is important for good quality decisions that the role of measurement variability be fully understood. One of the most important quality decisions is to determine if a lot conforms to the specification limits. This chapter discusses the fundamental concepts of the risks involved in making decisions based on data. The relationship of the concept of risk to specifications will be discussed in chapter 13 along with some common practices and related perils.

12.2 Fundamentals—The Decision Process

Any decision procedure can be described in terms of probabilities. Most decisions in life follow the same principles. The outcomes of a decision process relate to specifications as follows:

		The truth: Meets specifications	The truth: Does not meet specifications
The decision	Meets specifications	Correct decision	Incorrect decision (Customer's risk)
	Does not meet specifications	Incorrect decision (Producer's risk)	Correct decision

There are four outcomes. Two represent correct outcomes and two represent incorrect outcomes. The risks of incorrect decisions affect both the producer and the customer. The risks are directly related to the measurement variability and the amount of data used. Good practice calls for setting the risk levels as a business decision before selecting or designing sampling plans and acceptance procedures. The risks can never equal zero, but should be made small, such as 10 percent, 5 percent, or 1 percent.

12.3 Fundamentals—The Operating Characteristic Curve

In making decisions to release or accept product lots, the fourfold decision diagram always holds; however, the risks of correct and incorrect decisions vary with the level of measured characteristic. The performance of a release or acceptance plan is described by its operating characteristic (OC) curve. (See Bowker and Leiberman 1972.)

The plotted OC curve shows the probability of accepting the product for a range of *true* lot average values. It is developed from the sampling plan, the variability of the decision data, and the acceptance limit. The acceptance limit is the limiting value that the reported lot average may take and still meet requirements.

Figure 12.1 is an OC curve that shows the performance of an acceptance procedure for a maximum specification limit of 150 ppm benzene. It is an ideal OC curve, and shows what the performance would be if there were no measurement variability. With no measurement variability present the acceptance limit can be set at the specification limit, 150 ppm. The sampling plan is as follows:

- Take one sample per lot and measure it once.

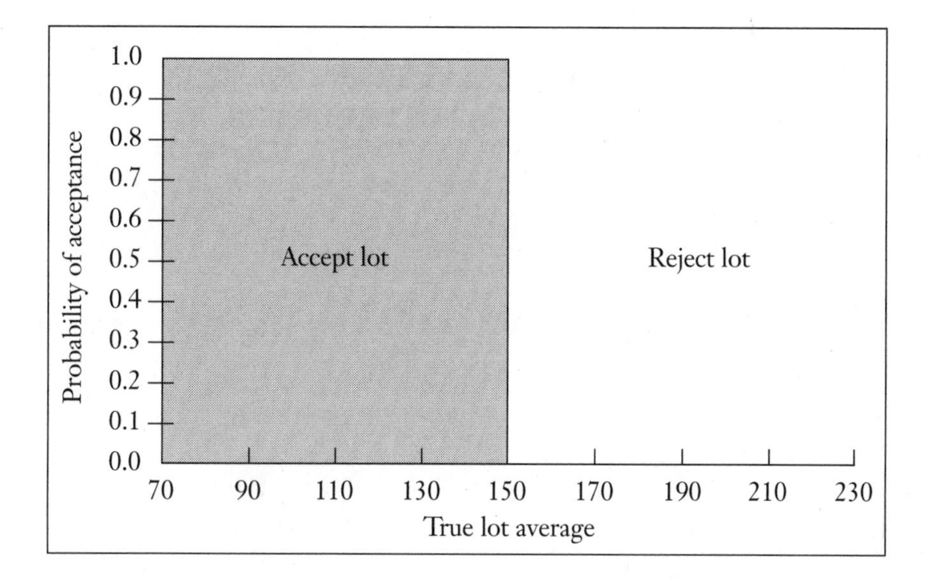

Figure 12.1. Ideal OC curve.

- Accept the lot, if test result is equal to or less than 150 ppm benzene.
- Reject the lot, if test result is greater than 150 ppm benzene.

Note that the acceptance limit and the specification limit are both set to 150 ppm. When the true product contains 150 ppm benzene or less, the probability of acceptance is 1.00. The product always is accepted. When the product has more than 150 ppm benzene, the product is always rejected.

In contrast to the ideal example, measurement variability is always present in actual operations. Thus, there is a degree of uncertainty in the decision to accept or reject a lot. The ideal curve does not describe the performance of real decision procedures. Figures 12.2, 12.3, and 12.4 show OC curves for a specification limit of 150 ppm and a standard deviation of the measurement variability of 10 ppm.

In Figure 12.2, the acceptance limit has arbitrarily been set at two times the measurement standard deviation *below* the specification limit. The acceptance limit is 130 ppm (150 – 20). Note that if the true lot

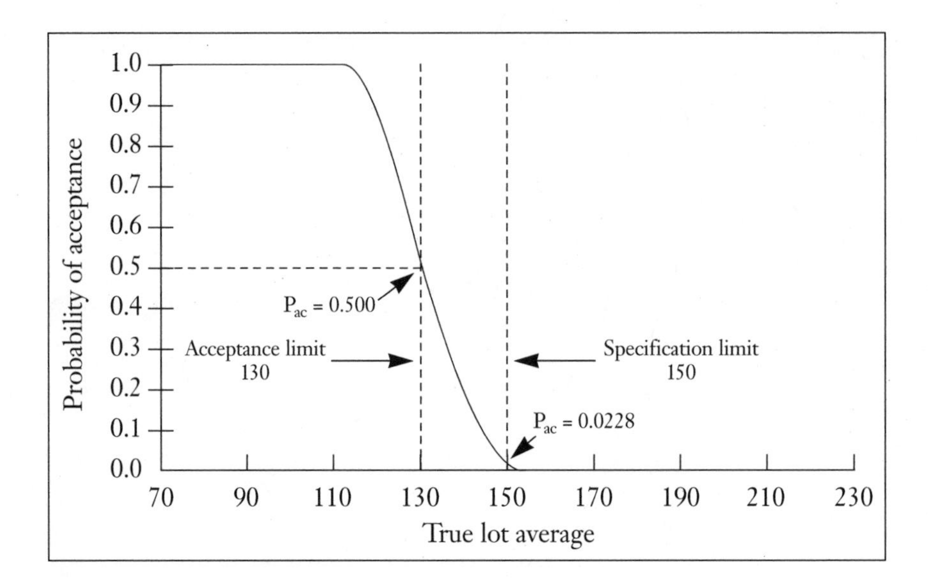

Figure 12.2. OC curve for measurement variability = s_{ind} = 10.

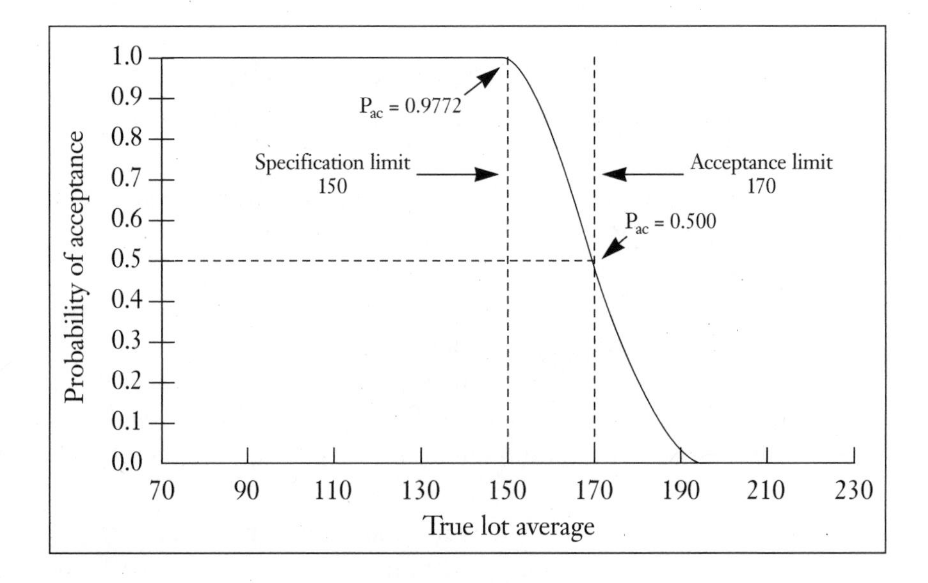

Figure 12.3. OC curve for two different sample sizes.

average is at the acceptance limit of 130 ppm, the probability of accepting the lot is 0.5 or 50 percent. If, however, the true lot average is at the specification limit of 150, the probability of acceptance is very small (0.0228). Many lots with averages between 130 ppm and 150 ppm that conform to the specification will be rejected. The plan described by this OC curve clearly protects the customer, but at a potential cost to the producer.

In Figure 12.3, the acceptance limit has been set at two times the measurement standard deviations *above* the specification limits. The acceptance limit is 170 ppm (150 + 20). Note that if the true lot average is at the acceptance limit of 170 ppm, the probability of accepting the lot is 0.5 or 50 percent. In contrast to the OC curve in Figure 12.2, the probability of acceptance when the true lot average is at the specification limit is very high (0.9772). Many lots with averages between 150 ppm and 190 ppm that do not conform to the specification lots will be accepted. This curve clearly protects the producer, but at a potential cost to the customer.

Clearly, there must be a balance between the acceptance limits and the specification limits so that the customer is protected and the costs to the producer are not prohibitive. One way to move the OC curve closer to the ideal curve is to base the decision on an average of more than one sample. This strategy reduces the standard deviation of the reported average. Figure 12.4 shows the OC curves for two acceptance procedures with different sample sizes.

	Curve 1	**Curve 2**
• Specification limit	150 ppm	150 ppm
• Sample size, n	1	4
• Standard deviation of reported average	10 ppm	5 ppm
• Acceptance limit	130 ppm	140 ppm

The acceptance limits for both OC curves have been set at two times the standard deviation of the reported result ($s_{avg} = s/\sqrt{n}$) below the specification limit. Therefore, both acceptance procedures have the same protection for the customer at the specification limit. But the OC curve for the acceptance plan with sample size of four shows better performance from the producer's point of view. Using the OC curves in this

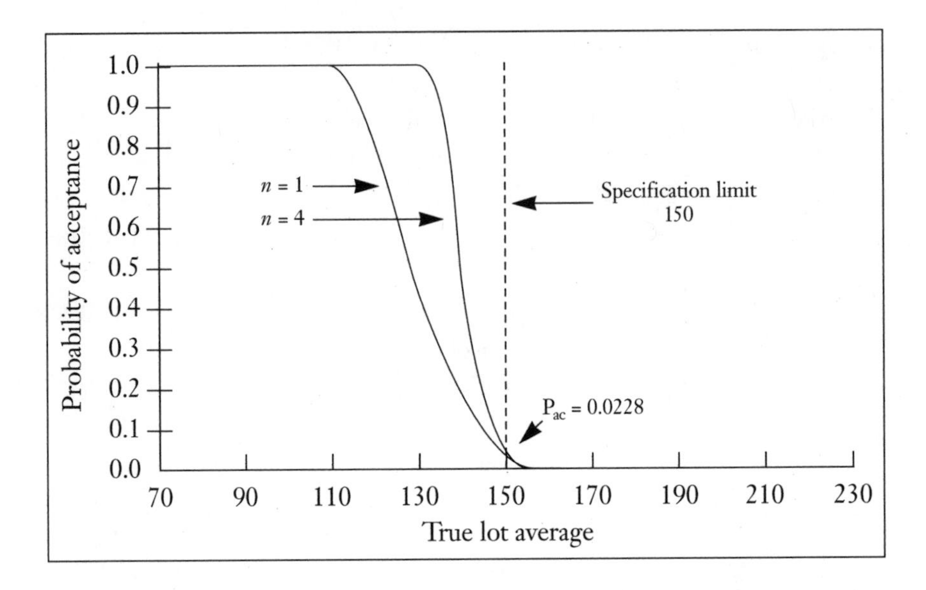

Figure 12.4. OC curves for two acceptance procedures with different sample sizes.

manner, the producer can balance the costs of increased sampling with the reduction in potential costs of rejecting conforming product.

The following chapter will provide more understanding of the OC curves and will show how they are developed.

12.4 The OC Curve and Specification Limits

Good quality practice calls for setting an acceptance limit inside the specification limit in order to provide a safety factor for the release decision. The OC curve provides a quantitative procedure to evaluate the risks involved in setting the safety factor interval between the specification limit and the acceptance limit. It is good practice to compute the OC curve for any proposed specification limit, acceptance limit, and sampling plan. It is often necessary to evaluate several configurations in order to find an economic plan that protects the customer at reasonable cost to the producer.

CHAPTER 13

Statistics for Decisions Based on Data

13.1 Introduction

A knowledge of basic statistical principles and concepts is necessary for effective decisions involving specification limits, acceptance or release limits, and sample size. These principles and concepts are in the body of knowledge for the ASQC Certified Quality Engineer program. The purpose of this chapter is to show how these limits, sample size, and the variability are related by the operating characteristic (OC) curve. (See Grubbs and Coon 1954.)

13.2 OC Curves—General

An OC curve is a graphical description of the performance of a decision process. OC curves are developed differently for attribute data than for variables data. The ASQC has publications for sampling procedures and tables for both attribute (ANSI/ASQC Z1.4-1993) and variables data (ANSI/ASQC Z1.9-1993). Both publications include the OC curves for the sampling plans. Other sets of published sampling plans also include OC curves.

The following discussion shows how OC curves are developed so that published curves can be interpreted correctly. It will also be useful when there is a need to plan for a specific application for which published sampling plans are not adequate.

13.3 OC Curves and Sampling Plans

The sampling plan and associated OC curve are defined by the following elements. The OC curve will be changed if any of these elements is changed.

- *Specification limits*—Limits that define the conformance boundaries for product characteristic of an individual unit of product. Setting specification limits is discussed in chapters 7 and 14.

- *Customer's risk*—The probability of accepting a lot when the quality of the lot has a designated numerical value representing a level at which it is seldom desirable to accept. It is the risk of accepting bad product. It is also called consumer's risk.

- *Producer's risk*—The probability of not accepting a lot when the quality of the lot has a designated numerical value representing a level at which it is generally desirable to accept. It is the risk of rejecting good product.

- *Variability of the data used for decision*—The change in the variability may result from a different sample size or from a change in the measurement process.

- *Sample size*—The number of observations in the average to be used for the acceptance decision.

An OC curve can be developed using these elements by determining the probability of acceptance for several levels of the true product average and plotting the results. Tables are commonly available so that detailed computation is not usually needed to determine the probabilities. The following sections describe basic OC curves for both attribute and variables data.

13.4 OC Curves for Attribute Data

An independent estimate of variability is not required for attribute data, because the variability is directly related to the observed count or percentage. In most cases a Poisson equation is usually adequate for attribute data, although at times a hypergeometric or binomial formula may be appropriate. (See Schilling 1982.)

As an example, consider the following elements.

- Specification limit 2 percent defective
- Customer's risk 10 percent (P_{ac} = 0.10) at 2 percent defective
- Producer's desire P_{re} = 0.90 at 1 percent defective
- Producer's risk 10 percent (P_{re} = .10) at 1 percent defective

Table 13.1 shows the probability of acceptance for sample sizes of 50, 100, 150, 200, and 400 and for acceptance limits of 0, 1, 2, 3, and 4 defectives per sample at the specification limit of 2 percent. The probabilities of acceptance have been taken from the tables of the cumulative Poisson distribution. The bold values are the combinations of sample size and acceptance limit that meet the customer's risk of 10 percent at a level of 2 percent defectives.

The next step is to select the sample size and acceptance limit that will minimize the producer's risk at a 1 percent defective level. Table 13.2 is similar to Table 13.1 but with a level of 1 percent defectives. The bold values are the combinations of sample size and acceptance limit that meet the desired producer's risk of 10 percent (P_{ac} = 0.90) at a level of 1 percent defectives.

A comparison of Tables 13.1 and 13.2 reveals that there is no combination of sample size and acceptance limit that will meet both the customer's and the producer's risks. The best acceptance plan for each

Table 13.1. Probability of acceptance at specification limit of 2 percent defectives.

Sample size	**Acceptance limit (maximum number of defectives in sample)**				
	0	**1**	**2**	**3**	**4**
50	.368	.736	.920	.981	.996
100	.135	.406	.677	.857	.947
150	**.050**	.199	.423	.647	.815
200	**.018**	**.092**	.238	.433	.629
400	**.000**	**.003**	**.014**	**.042**	**.100**

Table 13.2. Probability of acceptance at specification limit of 1 percent defectives.

Sample size	Acceptance limit (maximum number of defectives in sample)				
	0	1	2	3	4
50	.607	**.910**	**.986**	**.998**	**1.000**
100	.368	.736	**.920**	**.981**	**.996**
150	.223	.558	.809	**.934**	**.981**
200	.135	.406	.677	.857	**.947**
400	.018	.092	.238	.433	.629

sample size is the one that meets the customer's risk with minimum producer's risk. These plans are listed in Table 13.3.

No plan for sample sizes of 50 or 100 met the customer's risk of 10 percent. A large sample size is required to meet the customer's risk of 10 percent at the specification limit of 2 percent defectives. None of the plans will meet the producer's risk of 10 percent at 1 percent defectives. If this is unsatisfactory, a much larger sample size must be selected, new specification limits must be developed, or the risks must be increased.

If the customer's risk can be increased to 25 percent ($P_{ac} = 0.25$), the plans in Table 13.4 would meet the risk requirements.

Again, the producer's risk is not met. OC curves for the three plans that meet the customer's risk are shown in Figure 13.1 using data from the cumulative Poisson tables. Additional OC curves can be developed using different specification limits, acceptance limits, sample sizes, and risks until a sampling plan is found that is acceptable to both the producer and customer. In any case, the use of OC curves provides one of the best methods to evaluate the relationship between risks, limits, and sample size.

This example also illustrates that attribute sampling requires very large sample sizes. Variables sampling, in which characteristics are measured on a continuous scale, is used more frequently in the CPI. Smaller sample sizes can be used with variables sampling because variables data contain precise information about the characteristics.

Table 13.3. The best acceptance plans for each sample size at 10 percent customer's risk.

Sample size	Acceptance limit	Probability of acceptance at 1% defectives	at 2% defectives
150	0	.223	.050
200	1	.406	.092
400	4	.629	.100

Table 13.4. The best acceptance plans for each sample size at 25 percent customer's risk.

Sample size	Acceptance limit	Probability of acceptance at 1% defectives	at 2% defectives
100	0	.368	.135
150	1	.558	.199
200	2	.677	.238

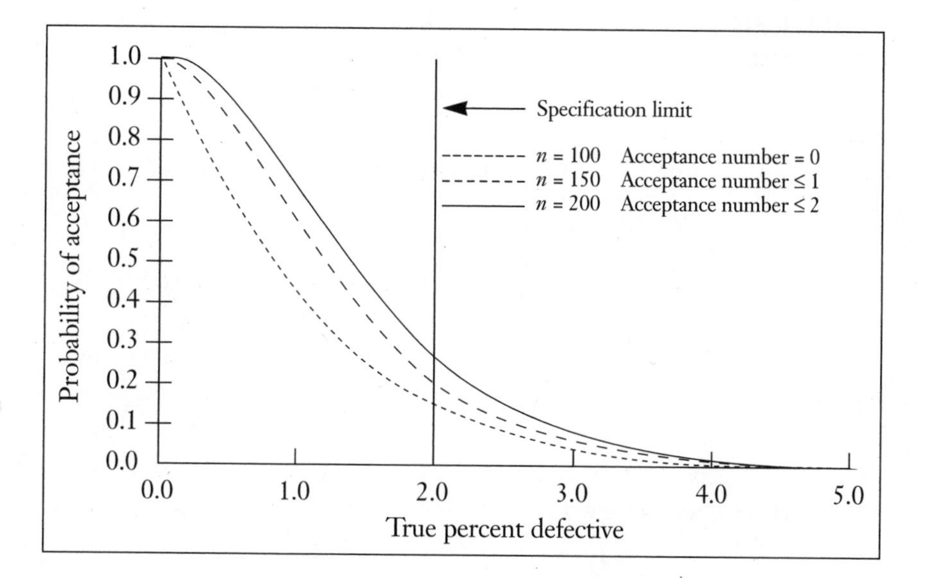

Figure 13.1. OC curve for three attribute sampling plans.

13.5 OC Curves for Variables Data

OC curves are calculated differently for variables data than for attribute data. For variables data, tables of the normal curve are used, and an independent estimate of the measurement variability is required. (See chapter 10.) Figure 13.2 shows the relationship between P_{ac} and the true lot average once an acceptance limit is specified. The five normal curves have lines dividing the distribution into standard deviation units. The standard deviation used is that for the average of the sample results used for the acceptance decision. The normal curves show the distribution of reported lot averages around each of the true lot averages.

$$s_{avg} = s_{ind} / \sqrt{n}$$

where s_{avg} is the standard deviation of the average
s_{ind} is the measurement standard deviation of individual results
n is the sample size (number of results in the average)

If the process average is at the maximum specification limit (bottom curve), an acceptance limit can be set to give a probability of acceptance equal to the customer's risk. For this example, the acceptance limit was set at two standard deviations from the specification limit.

If the sample average falls to the left of the acceptance limit, the product is accepted. The probability of acceptance is the portion of the area under the normal curve that falls within the acceptance zone, which is shown as a shaded area.

The area is found by computing

$$z = (\text{Acceptance limit} - \text{Process average}) / s_{avg}$$

For this case ($z = -2.000$), the normal table gives an area of 0.0228. It is shown as the probability of acceptance (P_{ac}). If the process average is 45 (second curve from bottom), which is one standard deviation less than the specification limit, then $z = -1.000$. This gives a P_{ac} of 0.1587, which is the area to the left of the acceptance limit. Similar calculations apply for the other process averages. If the process average is at the acceptance limit (middle curve), then $P_{ac} = 0.5000$, which means that the product

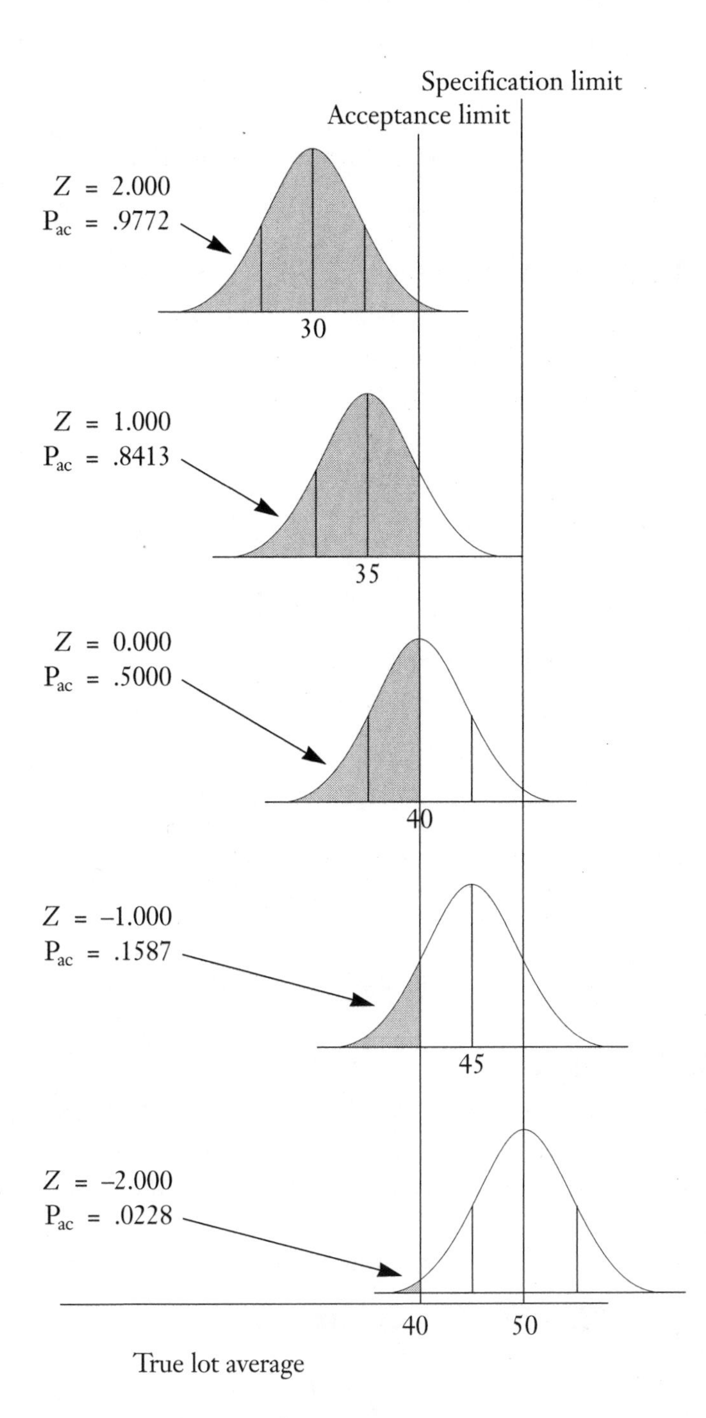

Figure 13.2. P_{ac} and the true lot average.

will be accepted half of the time and rejected half of the time. If the process average is below the acceptance limit (upper two curves), P_{ac} is greater than 0.5000. P_{ac} is plotted against the true process average to give the OC curve.

13.6 OC Curve—Setting the Acceptance Limit

An important application of the OC curve is to determine an acceptance limit given the specification limit, the customer's risk, the standard deviation of individual results, and the sample size.

Example: Specification limit = 50.0 (upper limit)
$n = 1$
$s_{ind} = 5.0$, therefore $s_{avg} = 5.0$
Customer's risk = 0.05

In the normal table, for $P_{ac} = 0.0500$, $z = -1.645$. Then,

$$\text{Acceptance limit} = \text{Specification limit} + z \times s_{avg}$$
$$= 50.0 + (-1.645)(5.0) = 41.8$$

On the other hand, if the specification limit of 50.0 were a lower limit, then

$$\text{Acceptance limit} = \text{Specification limit} - z \times s_{avg}$$
$$= 50.0 - (-1.645)(5.0) = 58.2$$

If there are both upper and lower specification limits, there will also be upper and lower acceptance limits.

13.7 Developing the OC Curve

The OC curve can then be developed by determining P_{ac} for several values of the process average covering the region of interest. Compute z for each value of the process average, and then use the normal curve tables to find P_{ac}.

For a maximum specification limit,

$$z = (\text{Upper acceptance limit} - \text{lot average}) / s_{avg}$$

For a minimum specification limit,

$$z = (\text{Lot average} - \text{Lower acceptance limit}) / s_{avg}$$

If the OC curve does not indicate a satisfactory producer's risk, additional OC curves can be developed for different sample sizes.

Table 13.5 shows the calculated values for OC curves using two different sample sizes.

Specification limit = 50.0 (upper limit)
Standard deviation of individual results = 5.0
Customer's risk = 5 percent (P_{ac} = 0.05, z = −1.645)

Table 13.5. Calculated values for OC curve example.

True lot average	Sample size = 1		Sample size = 4	
	Z	P_{ac}	Z	P_{ac}
20.0	4.36	1.0000	10.36	1.0000
22.0	3.96	0.9999	9.56	1.0000
24.0	3.56	0.9998	8.76	1.0000
26.0	3.16	0.9992	7.96	1.0000
28.0	2.76	0.9971	7.16	1.0000
30.0	2.36	0.9909	6.36	1.0000
32.0	1.96	0.9750	5.56	1.0000
34.0	1.56	0.9406	4.76	1.0000
36.0	1.16	0.8770	3.96	0.9999
38.0	0.76	0.7764	3.16	0.9992
40.0	0.36	0.6406	2.36	0.9909
42.0	−0.04	0.4840	1.56	0.9406
44.0	−0.44	0.3300	0.76	0.7764
46.0	−0.84	0.2005	−0.04	0.4840
48.0	−1.24	0.1075	−0.84	0.2005
50.0	−1.64	0.0505	−1.64	0.0505
52.0	−2.04	0.0207	−2.44	0.0073
54.0	−2.44	0.0073	−3.24	0.0006
56.0	−2.84	0.0023	−4.04	0.0001
58.0	−3.24	0.0006	−4.84	0.0000
60.0	−3.64	0.0001	−5.64	0.0000

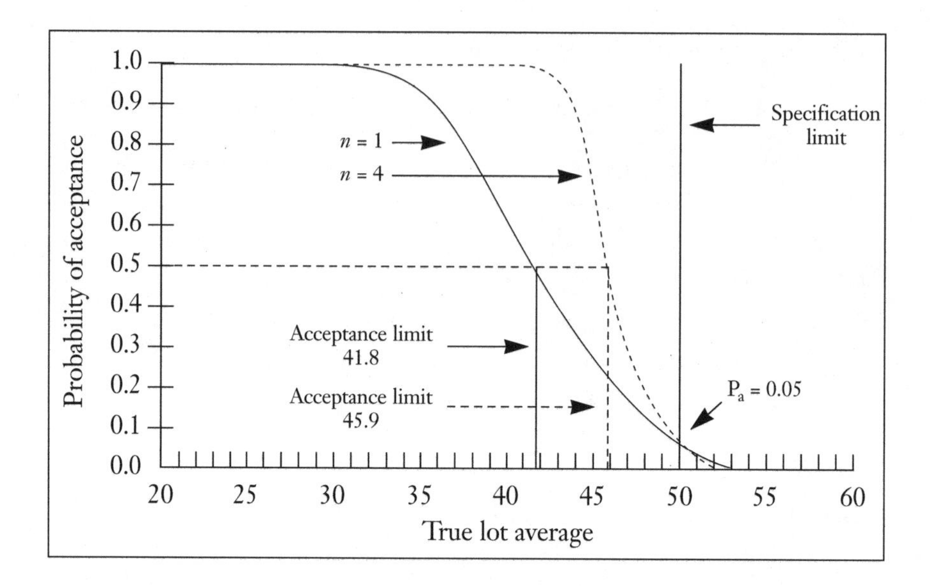

Figure 13.3. OC curve for two different sample sizes.

Sample size	S_{avg}	Acceptance limit	z
1	5.0	50.0 + (–1.645)(5.0) = 41.8	(41.8 – Lot avg) / 5.0
4	2.5	50.0 + (–1.645)(2.5) = 45.9	(45.9 – Lot avg) / 2.5

These OC curves are shown in Figure 13.3.

13.8 The Perils of Resampling

Resampling is the practice of taking another sample and using its results if the first sample result is rejected; that is, it is not within acceptance limits. This practice should be avoided. Double sampling is a better practice. (See section 13.9.) Resampling increases the probability of acceptance when the process is truly producing product within specification limits. It also, however, increases the probability of accepting product that is truly outside specification limits. This means that the customer's risk has increased from the level used to design the acceptance plan. This practice releases more product at the expense of the customer.

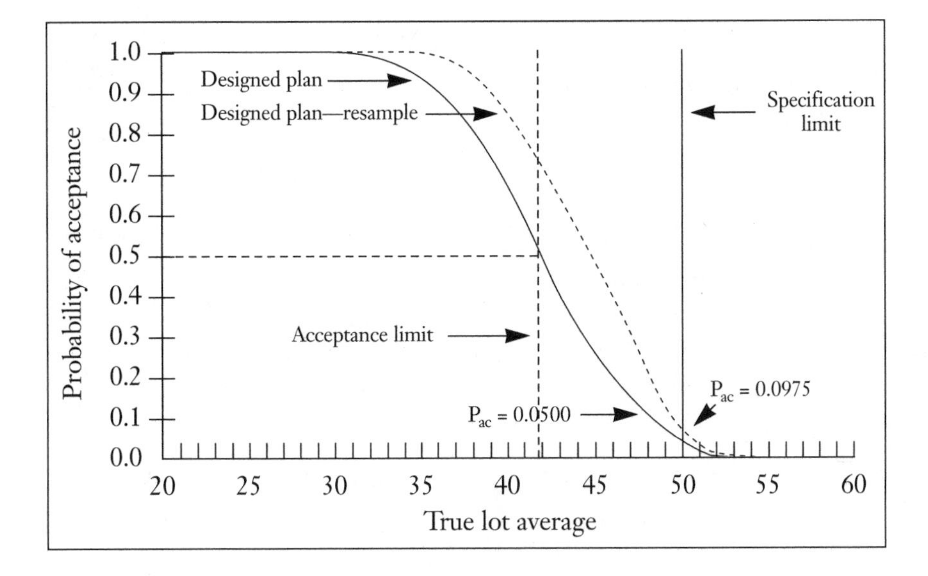

Figure 13.4. OC curve for resampling.

The probability of acceptance when resampling is done is

$$P_{ac}(\text{resample}) = P_{ac}(1) + P_{ac}(1)\,[1 - P_{ac}(1)]$$

$P_{ac}(1)$ is the probability of acceptance for the first sample computed as a single-sample plan.

If the customer's risk is 0.05 ($P_{ac}(1)$ at the specification limit), the P_{ac} (resample) is 0.05 + (0.05)(0.95) = 0.0975 or nearly double the risk used to design the acceptance plan. Figure 13.4 shows the OC curves for the designed acceptance plan with resampling.

13.9 Double Sampling

A better alternative to resampling is to use a double sampling plan. This increases the amount of product released and still protects the customer. A double sampling plan is operated as follows:

- Take a sample of size *n* (one or more).
- Average the results to obtain the reported lot average.
- If the reported lot average meets the acceptance limit, then accept the lot.

- If the reported lot average does not meet the acceptance limit, then
  - Collect another sample of size *n*.
  - Compute the reported lot average.
  - If this lot average meets the acceptance limit, accept the lot.
  - If this lot average does not meet the acceptance limit, reject the lot.

A double sampling plan can also be designed to compare the average of the first and second results with the acceptance limit. A double sampling plan using this average has a different acceptance limit for the same customer's risk than a double sampling plan that uses the individual values.

To design a double sampling plan, do the following:

1. Select the customer's risk (P_{ac}) as previously discussed (say 0.05).
2. Compute $P_{ac}(1)$ for the first sample at the specification limit.

$$P_{ac} = P_{ac}(1) + P_{ac}(1)[1 - P_{ac}(1)]$$

$$P_{ac} = 2P_{ac}(1) - P_{ac}(1)^2$$

$$\text{then } [P_{ac}(1)]^2 - 2P_{ac}(1) + P_{ac} = 0$$

3. Solve this quadratic equation.

$$P_{ac}(1) = \frac{2 - \sqrt{4 - 4P_{ac}}}{2} = 1 - \sqrt{1 - P_{ac}}$$

Let $P_{ac} = 0.05$ and solve for $P_{ac}(1)$.

$$P_{ac}(1) = 0.0253$$

4. Compute the acceptance limit.

$$z[(P_{ac}(1)] = -1.955 \text{ from normal curve tables}$$

$$\text{Acceptance limit} = \text{Specification limit} + z_{avg}$$

$$= 50 - (1.955)(5) = 40.2$$

Table 13.6 shows some of the calculations that would be used to generate the OC curve for double sampling.

Customer's risk = 0.05

Acceptance limit = 40.2

z = (Acceptance limit − Process average)/s_{avg}

$P_{ac}(1)$ = Probability of acceptance on the first sample

Table 13.6. Calculated values for OC curve example.

Lot average	Z	$P_{ac}(1)$	$P_{ac}(DS)$	
20.0	4.04	1.0000	1.0000	
25.0	3.04	0.9988	1.0000	
30.0	2.04	0.9793	0.9996	
35.0	1.04	0.8508	0.9777	
40.0	0.04	0.5160	0.7657	
45.0	–0.96	0.1685	0.3086	
50.0	–1.96	0.0250	0.0494	Specification limit
55.0	–2.96	0.0015	0.0030	
60.0	–3.96	0.0001	0.0002	

$P_{ac}(DS)$ = Probability of acceptance on the second sample

$P_{ac}(1) + P_{ac}(1)[1 - P_{ac}(1)]$

Figure 13.5 shows the OC curves for both the first and second samples of a double sample plan.

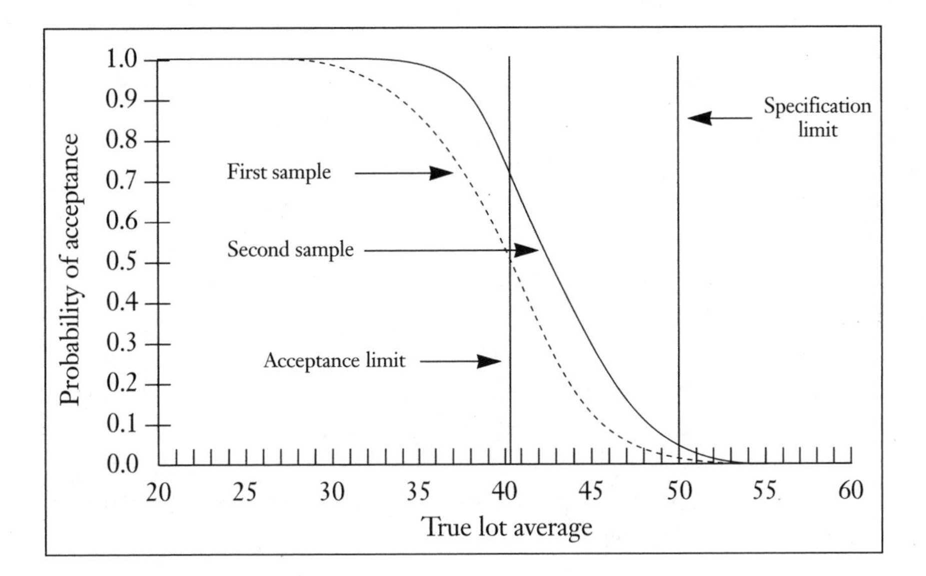

Figure 13.5. OC curve for double sampling plan.

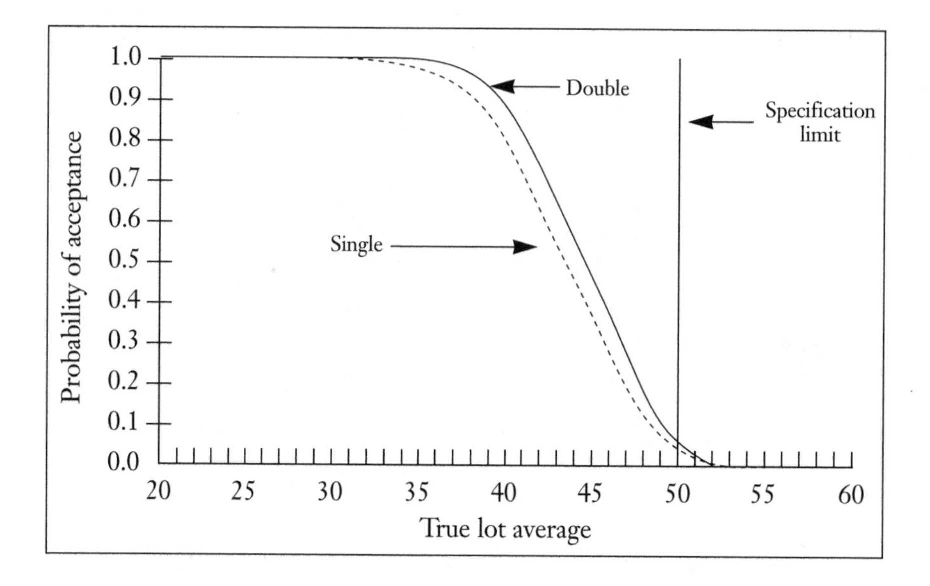

Figure 13.6. OC curve for single and double sample plans.

Figure 13.6 shows the OC curves for a single sample and double sample plan giving the same customer's risk.

13.10 An Analytical Solution to Determine Sample Size

The specification limits apply to individual units of product. The acceptance limits are limits on the reported lot average. If the variability within the lot is large, then it is necessary to find the values for the maximum and minimum lot averages so that individual units will conform to the specification limits. (See Guttman, Wilks, and Hunter 1965.)

The specification limit and the customer's risk must be defined when designing an acceptance plan. Sometimes the producer's risk at a certain lot average is defined through either needs or desires. Using this information, it is possible to determine by direct solution the sample size that will meet both the producer's and customer's risks. Once the sample size has been determined, the acceptance limit can also be determined using the procedures discussed.

Example:

Specification limit = 50.0

Customer's risk = 0.05

Producer's risk = 0.10 at true lot average = 45.0

Standard deviation s_{ind} = 5.0 of individuals

Therefore, P_{ac} at 50 = 0.05 z(Customer's risk) = –1.645

P_{ac} at 40 = 0.90 z(Producer's risk) = 1.282

The acceptance limit must be higher than the producer's acceptable lot average and lower than the specification limit.

The following equation is needed for each of the two lot averages 40 and 50.

$$\frac{[\text{Acceptance limit} - \text{Lot average}]}{s_{ind}/\sqrt{n}} = z$$

At the producer's limit,
(Acceptance limit – 40.0)/(5.0/$\sqrt{n}$) = 1.282

At the specification limit,
(Acceptance limit – 50.0)/(5.0/$\sqrt{n}$) = 1.645

Solving these equations for sample size (n) reduces to

$$\sqrt{n} = \frac{[(s_{ind})(z_p) - (s_{ind})(z_p)]}{[LA_s - LA_p]}$$

where LA_s is lot average at the specification limit
LA_p is lot average at producer's limit

$$\sqrt{n} = \frac{[(5)(1.282) - (5)(-1.645)]}{[50 - 40]} = 1.464$$

n = 2.14 (round up to 3)

Solving both equations for the acceptance limit yields the following:

Using the producer's limit,

$$\text{Acceptance limit} = (1.282)(5.0) / 2 + 40.0 = 43.7$$

Using the specification limit,

$$\text{Acceptance limit} = (-2.575)(5.0) / 2 + 50.0 = 45.3$$

The two equations do not agree exactly because the sample size was rounded from 2.14 to 3. Use the average of the two results or 44.5—the acceptance limit. The probability of acceptance at the producer's limit of 40.0 is $z = (44.5 - 40.0)/(5.0/1.732) = 1.559$; therefore, $P_{ac} = 0.9405$. The probability of acceptance at the specification limit of 50.0 is $z = (44.5 - 50.0)/(5.0/1.732) = -1.905$; therefore, $P_{ac} = 0.0289$. The requirements are met.

CHAPTER 14

Negotiating Specification Limits

14.1 Introduction

The process of setting specification limits always implies a negotiation. The negotiation takes place between the producer and the customer. The producer and customer are usually thought of as representing different companies; however, they may represent different units within a single company in the case of negotiating internal specifications. The negotiation may range from active bargaining across a table to the simple agreement to the proposal by one of the parties. The content of the negotiation depends upon the process knowledge that each party brings to the proceedings.

- The producer should know the average value of each characteristic of concern, product variability, and measurement variability. (See chapter 9.) In addition, the producer should know whether the process is in a state of statistical control and if the process target can be changed.

- The customer should know the desired target value, the tolerable variability about the target for each characteristic of interest, what compromises among characteristics are acceptable, and what packaged quantity is required to meet the specification limits.

Where practical, an exchange of process knowledge and understanding will facilitate negotiations. (See chapter 21.)

14.2 Negotiations

Several scenarios for the negotiation are possible. These are summarized in Figure 14.1. The two most common scenarios occur when the customer requests specification limits and when the customer does not request specification limits.

14.2.1 The Customer Requests Specification Limits

Often the customer requests specification limits. This scenario also applies for specification limits set by government or industry groups. The producer has four options.

1. If the process is in statistical control and capable ($C_{pk} \geq 1.33$), the customer's request can be met. (See chapter 15.) The producer can accept the customer's requested specification limits.

2. If the process is not in statistical control but past performance lies within the customer's requested limits ($P_{pk} \geq 1.33$), the producer may accept the proposed limits with the risk that the process must be monitored and each lot must be tested for conformance. The producer should assess the performance of any proposed release procedure with an OC curve. (See chapter 13.) The producer should attempt to identify and remove special causes and bring the process into a state of control.

3. If the process is in statistical control but not capable ($C_{pk} < 1.33$), the producer has four choices.

   a. Accept the proposed limits with the added complexity of having to sort lots to meet the customer's specification. The producer should assess the performance of any proposed sorting release procedure with an OC curve. (See chapters 13 and 16.)

   b. Accept the proposed limits with the commitment to implement physical changes in the system to reduce the process variability. This may require defining a new product.

   c. Propose wider specification limits. Good practice calls for specification limits at $T \pm 4s_{TOT}$, where T is the target value and s is the standard deviation. (This value sets the initial P_p at 1.33.) Whether to use the total or product standard deviations, s_{TOT} or s_{PROD}, is a judgment call. If the s_{TOT} is much larger

Customer requests specification limits		
Product meets limits?	**Process in control?**	**Producer's action(s)**
Yes	**Yes**	Producer can accept limits.
Yes	**No**	Producer can accept limits with risk, but needs to • Monitor process. • Assess release procedure (OC curve). • Test each lot. • Work toward control.
No	**Yes**	Producer has the following choices. a. Accept limits, assess release procedure, sort lots to meet customer's limits. b. Accept limits with commitment to implement changes in system. c. Propose wider specification limits. d. Decline the business.
No	**No**	Producer should not accept limits until changes are made in the system to bring the process in control.

Customer does not request specification limits	
Process in control?	**Producer's action(s)**
Yes	Producer can propose limits at $\pm 4s_{WLS}$.
No	Producer can determine and negotiate temporary limits. a. Plot and examine the available data (time plots and histograms). b. Estimate minimum and maximum values for X (actual minimum and maximum if $n > 50$). c. Estimate measurement variability (s_{MEAS}). d. Propose temporary limits at $X_{MIN} - 3s_{MEAS}$ and $X_{MAX} + 3s_{MEAS}$. e. Monitor process. Test each lot for conformance to the temporary limits. f. Work toward control. g. Plan for negotiation of permanent limits.

Figure 14.1. Negotiation scenarios.

than s_{PROD}, then measurement is a major contributor the to the variability and s_{PROD} may be used for the estimate.

d. Decline the business.

4. If the process is not in control and past performance does not lie within the customer's requested limits, then the producer should not accept limits until changes are made in the system to bring the process in control. The producer may accept the proposed limits with the risk that the process must be monitored and each lot must be tested for conformance to the specification limits. The only assurance that the product conforms to the customer's requirements is the test and release system. The issues of OC curves (chapter 13) and sorting (chapter 17) also must be considered.

14.2.2 The Customer Does Not Request Specification Limits

When the customer does not request specific specification limits, or if the producer is developing a new product, the producer has two options.

1. If the process is in control and capable, good practice calls for setting specification limits at $T \pm 4\sigma_{WLS}$, where σ_{WLS} is the standard deviation of within-lot product and measurement away from the target T. (This value establishes the initial C_p at 1.33.)

2. If the process is not in control and temporary specification limits are required for business reasons, the producer must carefully assess the data available to determine and negotiate temporary specification limits. Good practice calls for the following actions.

   a. Plot and examine the available data as time plots and histograms. Examine the plot for shifts, trends, and clustering. Screen for outliers that can be attributed to special causes.

   b. If the sample is large (>50), use the actual minimum and maximum. If the sample is small, estimate the minimum X_{MIN} and maximum X_{MAX} values based on the best technical judgment available.

   c. Estimate the measurement variability. Compute a standard deviation for measurement and sampling s_{MEAS} from replicate sampling or from a variance component analysis.

d. Set the temporary specification limits at $X_{MIN} - 3s_{MEAS}$ and $X_{MAX} + 3s_{MEAS}$.

e. Monitor the process to verify that the process stays on target and that the variability will stay constant. If the process is not in control these characteristics cannot be guaranteed.

f. Test each lot for conformance to the temporary specification limits. The only assurance that the product conforms to these specification limits is the test and release system. The issues of OC curves (chapter 13) and sorting (chapter 17) must also be considered.

g. Plan for the negotiation of permanent specification limits. The temporary specification limits allow for the introduction of the product. As more data are obtained and as the better process control is achieved, permanent specification limits can be established.

14.3 Special Considerations for One-Sided Specification Limits

In chapter 7 it was pointed out that some characteristics require only a single specification limit because variation on the other side is overcome by nature or measurement detection. A one-sided specification limit is often set before the target is determined. In such cases it is important to define the target relative the specification limit. The principle applies in reverse. The target should be set at $4\sigma_{TOT}$ inside the specification limit. This guideline is consistent with the use of an individuals control chart for the characteristics.

CHAPTER 15

Capability and Performance Measures

15.1 Overview

There has always been a need to quantify the past performance of a process as well as to estimate its future performance. The old measure, yield, has limitations. Inspection yield is computed as $Y = A/B$, where A is the amount of accepted product and B is the amount of product submitted for inspection. While this statistic has an immediate economic impact for managers, it is unreliable as a measure of quality for several reasons.

- The ratio of the number of lots accepted versus the number of lots submitted for inspection, A/B, gives no clue for corrective action.
- The amounts in the calculation of inspection yield may be determined in different ways; for example, by weight, volume, units, and so on.
- The effect of rework, subgrade, upgrade, and special assignments may subvert the intent of the measure.

Two useful measures used extensively throughout this book are process capability indexes and process performance indexes.

15.2 Process Capability

Process capability is a statistical measure of the inherent process variability for a given characteristic. Process capability for continuous characteristics is defined as 6σ, where σ is the standard deviation of the process in a state of statistical control. This value of σ is equivalent to σ_{WLS} as defined in section 10.7. When a process is in a state of statistical control the long-term components of variance are negligible. This value is often estimated by a range (or moving range) chart. (See Ryan 1989.)

Capability describes only the process variability when the process is operating at its best—in a state of statistical control. Process capability is measured in the units of the characteristic and, as such, cannot be used to compare different characteristics. It also does not include any indication of control to target.

15.3 Capability Indexes

The capability indexes that are in wide use today involve assumptions that are often unrecognized or ignored. Each index has advantages and disadvantages. No one index fulfills all needs. (See Chrysler Corporation, Ford Motor Company, and General Motors Corporation 1991.)

Capability indexes are dimensionless numbers that combine specification limits and a current estimate of process capability. Because they are dimensionless, it is widely believed that process indexes allow comparison between different producers, different products, different processes, and even different industries. Such comparisons are often not valid because different calculation procedures have been used by different producers.

15.4 Index Measures in Relation to Specification Limits

Four capability indexes are in common use. All capability indexes require the process to be in a state of statistical control. If the process is in statistical control, then the distribution of individual observations from the process will follow the normal distribution (or they can be transformed into the normal distribution). Thus, the statistics can be used to predict future operation of the process.

- $C_p = (USL - LSL) / 6\sigma_{WLS}$

where USL is the upper specification limit

LSL is the lower specification limit

σ_{WLS} is the current within-lot product and measurement standard deviation. σ_{WLS} may be estimated from variance components or from a control chart showing the process to be in control.

C_p is the basic capability index. It is a ratio of the specification range divided by the current process capability. C_p assumes a two-sided specification. To be valid for comparison with other processes, the specification limits must have been set to express the same original, relative process spread. It also assumes that the process distribution is normal and the process centered exactly on target (Kane 1986).

According to *Quality Assurance for the Chemical and Process Industries—A Manual of Good Practices* (1987, 37),

> Values of C_p exceeding 1.33 indicate that the process is adequate to meet the specifications. Values of C_p between 1.33 and 1.00 indicate that the process, while adequate to meet the specifications, will require close control. Values of C_p below 1.00 indicate the process is not adequate to meet the specifications and that the process and/or specifications must be changed.

If the original specification spread has been set to express a range of $\pm 3\sigma_{WLS}$, then the C_p cannot be expected to exceed 1.00 until substantial reduction in process variation has occurred.

- $C_{pk} = \text{Minimum } \{[(USL - P_A) / 3\sigma_{WLS}], [(P_A - LSL) / 3\sigma_{WLS}]\}$

where P_A is the current process average.

C_{pk} is the capability index adjusted for location. It measures whether the process is capable of meeting the requirements by considering the specification range, the current process average P_A, and the current σ, assuming normality. C_{pk} is applicable for either two-sided or one-sided specification limits. If the process is not centered, then the C_{pk} will be lower than C_p.

C_{pk} = 1.00 indicates that the process is capable of meeting specifications, provided that the process mean does not change.

C_{pk} = 1.33 indicates that the process is capable of meeting specifications, even if the process mean changes by up to $1\sigma_{WLS}$ from the target

- $C_{pm} = (USL - LSL) / \{6\sqrt{(P_A - T)^2 + \sigma_{WLS}^2}\}$

where T is the process target.

C_{pm} is the capability index relative to the target. The denominator includes the root-mean-square deviation from the target and is consistent with the Taguchi quality loss function. The target need not be in the center of the specification range for this measure. C_{pm} also has the same expectation restrictions as stated for C_{pk}.

- $C_t = (USL - LSL) / 6\sigma_{ST}$

where σ_{ST} is the current standard deviation for short-term measurement and sampling only. The value of σ_{ST} may be estimated from variance components or from a control chart on the measurement process to separate it from σ_{TOT}. (See chapter 10.)

C_t is the capability of the measurement process in a state of control. This index assumes a two-sided specification. It also assumes that the measurement process distribution is normal and that measurement bias is negligible. If the measurement variance component ranges from 10 percent to 60 percent of the total variance, then C_t will range from 5.00 to 2.00.

Other capability indexes can and have been defined to meet special requirements. The user should take time to understand special indexes that may be proposed for special situations. A new standard, BSR/ASQC Z1.10, *Standard Method for Calculating Process Capability and Performance Measures,* is in preparation.

Good practice calls for reporting capability indexes only for processes that are in statistical control. If an index is required for a process not in statistical control a performance index should be used. (See section 15.5.)

C_{pk} is preferred to C_p because it also includes the process location. The C_{pm} index is often preferred because the variability used in computing the index is consistent with the Taguchi loss function. (See section 7.2.)

A single index is not sufficient to describe a process. A comparison of C_{pk} (or C_{pm}) and C_p can be used to indicate if high variability or a bias from the target is a problem.

15.5 Process Performance

Process performance should be distinguished from process capability. Process performance represents the actual distribution of product and measurement variability over a long period of time, such as weeks or months. On the other hand, process capability represents the product and process variability over a short period such as minutes, hours, or a few batches. The variation of process performance will have a wider spread than process capability variation, because process performance includes additional components of variance due to time factors. Performance indexes analogous to the capability indexes can be defined as follows:

- $P_p = (USL - LSL) / 6s_{TOT}$
- $P_{pk} = \text{Min}\,\{[(USL - P_A) / 3s_{TOT}], [(P_A - LSL) / 3s_{TOT}]\}$
- $P_{pm} = (USL - LSL) / \{6\sqrt{[(P_A - T)^2 + s^2_{TOT}]}\}$
- $P_t = (USL - LSL) / 6s_{MEAS}$

The sample standard deviation for the extended period of interest is s_{TOT}. The measurement standard deviation is s_{MEAS}. They may be estimated from variance components or from direct computation using a representative sample for the period in question. (See chapter 10.)

Some users find that a performance index based on only the product components of variance is helpful in communicating the role of measurement and sampling variability. This index may be designated as P_{prod} and is defined as $P_{prod} = (USL - LSL) / 6\sqrt{[s^2_{LOT} + s^2_{WL}]}$.

If P_{prod} is much smaller than P_p, then measurement and sampling are the major sources of variability and the product is identified as relatively uniform. Such knowledge can prevent wasted effort of the producer in reducing the variability of a test method that does not benefit the customer.

Performance indexes are also subject to the same expectations as expressed for C_p. If P_t is much larger than C_t, then the long-term measurement component is a large part of the measurement process variability.

15.6 Variability of Capability and Performance Indexes

Anyone using capability or performance indexes should be aware of the variability in such indexes. Table 15.1 gives 95 percent and Table 15.2 gives 99 percent confidence limits for P_{pk} for levels of 1.00, 1.33, and 1.67. These values have been determined by 100,000 simulations each (Kittlitz 1987). Although computed for P_{pk}, these limits are appropriate for the other indexes as well.

Note that these confidence intervals are not only wider than many users would like to see, but they are not symmetrical about the nominal value.

Table 15.1. 95 percent confidence interval for P_{pk}.

n	P_{pk} = 1.00		P_{pk} = 1.33		P_{pk} = 1.67	
30	0.76	1.31	1.02	1.76	1.29	2.19
60	0.83	1.21	1.11	1.61	1.49	2.01
120	0.88	1.14	1.17	1.52	1.47	1.90

Table 15.2. 99 percent confidence interval for P_{pk}.

n	P_{pk} = 1.00		P_{pk} = 1.33		P_{pk} = 1.67	
30	0.71	1.47	0.95	1.95	1.20	2.43
60	0.78	1.30	1.05	1.71	1.32	2.15
120	0.84	1.20	1.13	1.59	1.42	1.99

15.7 PPM Reporting

Several CPI companies express the fraction nonconforming units resulting from their processes as a percentage. As quality improvement programs achieve practical results, the fraction nonconforming will be reduced. It can also be expressed as parts per million (ppm). Management is sometimes interested in using these metrics in a capability index or performance index sense. Some business leaders have set goals of achieving world-class critical levels of nonconforming units in the 0 to 500 ppm range.

An attractive feature of a parts-per-million (PPM) index is that both manufacturing and service characteristics can be assessed with the same index. For example, the ppm of defective product can be compared with the ppm of defective invoices. These measures can be used to assist the producer in the prioritization of process improvement efforts across functions.

Tables for the two-tail area of a normal distribution z are often used to determine the ppm nonconforming value corresponding to a C_p value. Enter the table with $z = 3C_p$. For example, if $C_p = 1.00$, the value of $z = 3.00$ yields a fraction nonconforming 0.0027, which may be expressed as 2700 ppm.

PPM statistics may be used in two ways.

1. In the *process capability* sense, all of the restrictions of a C_p index apply because the measure is considered a prediction of future performance. Specifically, the underlying population must be normal (or transformable to normal), and the process must be stable.

2. In a *process performance* sense, the conditions of a P_p index apply. Neither a normal population nor a stable process is required; however, an adequate sample of historical data to represent a meaningful period of operation is required. Thus, the ppm value reported is a practical measure of historical performance, but it cannot be used to predict future performance.

The user of PPM indexes is cautioned to be aware that

- They may not be additive. Most products have multiple characteristics, all of which must be in conformance for the product unit to be in conformance. C_p and P_p indexes are most often used on individual characteristics. A PPM index is often used in relation to the total fraction nonconforming, which may have more than one type of nonconformity. Unless the characteristics are mutually exclusive, the sum of the individual ppm nonconforming for multiple characteristics may exceed the total ppm nonconforming.

• Product units may be vaguely defined. Chemical and physical characteristics are often reported on the average characteristic level for a batch or lot. The reported value does not represent individual product units. Thus, the PPM index does not represent the number of defective units. It represents the probability of failing to accept the lot.

The standard deviation of the reported values may be based in large part on the measurement components of variance. (See chapter 10.) The portion of the distribution beyond specification limits does not estimate the percentage (or number) of defective units since there are no units. The portion beyond specification limits expresses the probability that the reported lot average may be beyond specification limits and that the entire batch may be judged nonconforming. Thus, the ppm = 2700 for a process with a $C_p = 1.00$ expresses the probability of failing to accept the lot. That is, there are 2700 chances in a million that the lot fails even when it is perfectly centered on target.

• The attempt to reduce complex systems to a single number inevitably obscures important information that may be required to understand and improve the system. The index by itself cannot improve quality.

15.8 Six-Sigma Goal

The six-sigma goal often stressed in total quality seminars is really expressing a need for a 12-sigma specification range. This is equivalent to a C_p of 2.0 if the process is on target, or a C_{pk} of 1.5 if the process is 1.5 sigma off target. These levels may be difficult or impossible to attain in systems with large measurement variability. The cautions cited in section 15.7 also apply to the six-sigma goal. If specification limits can be widened to express the actual requirements of the customer's process, the six-sigma goal may be feasible in the CPI. Agreement to widen the specification limits, however they were originally set, is very difficult to obtain from the customer, and is a poor quality practice.

This is not to say that reduction in variability cannot be made or should not be sought. Reducing the measurement product variability will often reduce the total variability with less effort than reducing process variability. Process improvement to reduce variability, however, is of real benefit to the customer, and should be the focus of quality improvement programs. Reduction in both process and measurement variability may be attainable through use of design of experiments and other optimization techniques. (See Box, Hunter, and Hunter 1978.)

Part IV

Using Specifications

CHAPTER 16

Conformance Decisions

16.1 Conformance Decisions and Diagram

After the specifications have been developed and approved, periodic checks are necessary to ensure that the process and product conform to the specifications. If the process or product does not conform to the specifications, then decisions must be made to determine the disposition of the nonconforming product. Figure 16.1 is a decision diagram that shows the relationships among the following:

- The factors used to define specifications and the methods used to ensure conformance
- The elements for evaluation of the process and product
- The disposition decisions for the product
- The handling of nonconforming product

It is important to establish procedures for handling nonconforming product. The goal, of course, is to continually improve the process so that the process always conforms to the specifications.

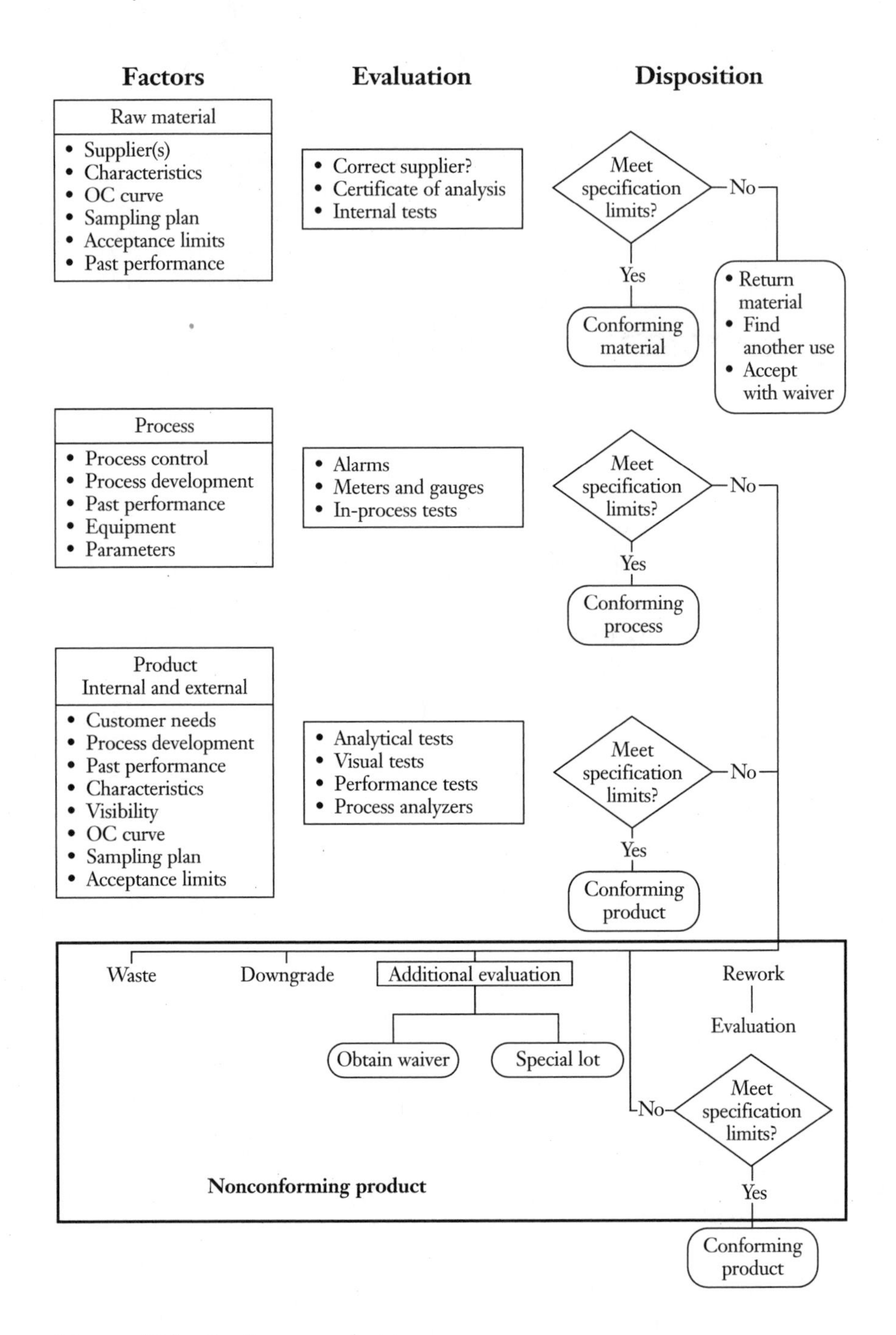

Figure 16.1. Conformance decisions.

16.2 Nonconforming Product

Nonconforming product is a unit of product that contains one or more nonconformities. In other words, one or more product characteristics or process parameters fail to meet the specification.

It is never a good quality practice to ship nonconforming product without informed agreement from the customer to accept the product.

A decision that product is nonconforming does not mean that the product does not meet the customer's requirements. It means that one or more of the following conditions exist and should be corrected.

- A raw material did not meet the raw material specification.
- A process parameter did not meet the process specification.
- One or more product characteristics did not meet the product specification.
- A piece of test equipment (on-line instrumentation or laboratory) could not measure a process parameter or a product characteristic.
- The test equipment was not calibrated.
- The sampling, testing or acceptance procedures were not performed as specified.

Further evaluation and testing is often necessary to make valid decisions for the disposition of nonconforming product.

16.3 Handling Nonconforming Product

Nonconforming product must be segregated from conforming product. Procedures should be in place that define the necessary steps for evaluation and disposition of the segregated product. (See the ANSI/ISO/ASQC Q9000 standards.) The proper disposition can be dependent on the type of nonconformity. Some common practices are

- *Waste*—The product may be disposed of as waste if it is not acceptable for another use.

- *Change grade*—The product may be usable as a product that meets a different specification. This action may also result in a lower sales price.

- *Special lot*—The product may be usable for the same end use as the intended product; however, the product is handled separately and is not mixed with conforming product. A waiver is usually required. There may or may not be a sales price penalty. (See chapter 19.)

- *Rework*—The nonconformity may be removed by reworking the product. The process of reworking, however, must not cause any of the other product characteristics to move out of specification limits. This must be verified.

CHAPTER 17

Product Sorting

17.1 Grading and Sorting

Grade is a specific category or range of characteristics of a product. Grade reflects a planned difference in requirements. Ideally, a process should produce product with a sufficiently narrow spread in the characteristic that all product produced is within the specification limits.

If, however, the process variability is much greater than that required to meet specifications, or if the process cannot produce on target, then sorting may be required to produce the appropriate grades. Sorting into grades is a result of not achieving needed process consistency, and carries the risks and problems inherent in lack of product uniformity. Therefore, sorting into grades is not a good quality practice.

The production of multiple coproducts from one feedstock is an appropriate use of grades. An example is fractional distillation of hydrocarbons. The "tops" are high-octane gasoline, the "middle" cuts are lube oils, and the "bottoms" are pitches. A contrary example is polymer viscosity. Consider a polymer that is sold in several narrow viscosity ranges. These ranges are grades. If these narrow grades are produced by making the product and then determining—by inspection or test—the lots or units of material falling into each grade, then the grades have been achieved by sorting.

17.2 Specifications for Product Grades

Each grade or category should have specification limits that appropriately define that grade. If a customer can receive product from more than one category, it is acceptable to have one product specification that shows the different specification limits for each category.

17.3 Product Identification for Graded and Sorted Product

It is not good quality practice to mix product from different categories. For example, if the product is separated into three categories (low, middle, and high), there are at least five possible product distributions that a customer may receive.

- Unsorted product
- Low sort
- Mid sort (sometimes called heart cut)
- High sort
- Product with heart cut removed

Each sorted category has a unique distribution and should have unique identification and specification limits.

17.4 The Effect of Measurement Variability on Sorting

When measurement variation is present, grading and sorting do not separate product as precisely as desired. The measurement variance often can be from 10 percent to 60 percent of the total observed variance.

Table 17.1 shows the values used to calculate Figures 17.1–17.4. These figures show the effect of sorting product in the presence of measurement variability.

- Figures 17.1 and 17.3 show the effect of grading product when the measurement variance is 10 percent of the total variance.
- Figures 17.2 and 17.4 show the effect when measurement variance is 60 percent of the total variance.
- Figures 17.1 and 17.2 show the individual product distributions in the low, mid, and high categories.

- Figures 17.3 and 17.4 show the distributions for the total product and with the mid or heart cut removed.

Tables 17.1 and 17.2 summarize the calculations for Figures 17.1–17.4. Note that the product variance and the acceptance limits are the same for both cases. Figures 17.1 and 17.2 show that

- There is only one target for the three groups.
- There is considerable overlap in the distributions of the three categories.
- When measurement variability increases, the differences among category averages decrease, and the variability within the categories increases.

Figures 17.3 and 17.4 show the following:

- The product with the mid category or heart cut removed is the combination of the low and high categories, and has the same range of values as the total product.
- When measurement variability is small, the differences between the total product and the product with heart cut removed are much greater than when measurement variability is large.

Each of the five product distributions identified in section 17.3 is distinctly different and should be handled as a different product. There will probably be differences in the performance in the customer's processes among the five products.

Table 17.1. Values used in the calculations for Figures 17.1–17.4.

	Figures 17.1 and 17.3	**Figures 17.2 and 17.4**
Percent measurement variance	10%	60%
Process average	100	100
Product variance	18	18
Measurement variance	2	27
Total observed variance	20	45

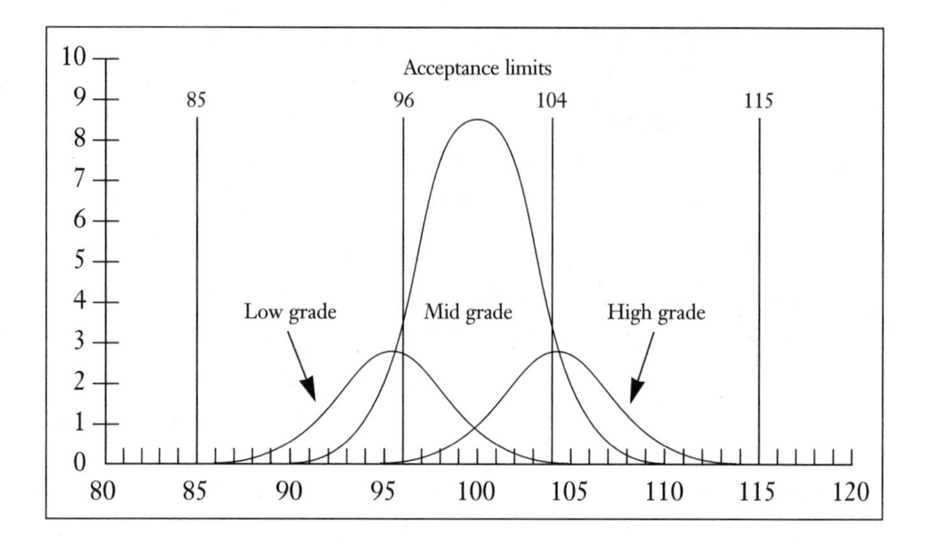

Figure 17.1. Product distribution by grading/sorting: Measurement variance = 10 percent of total variance.

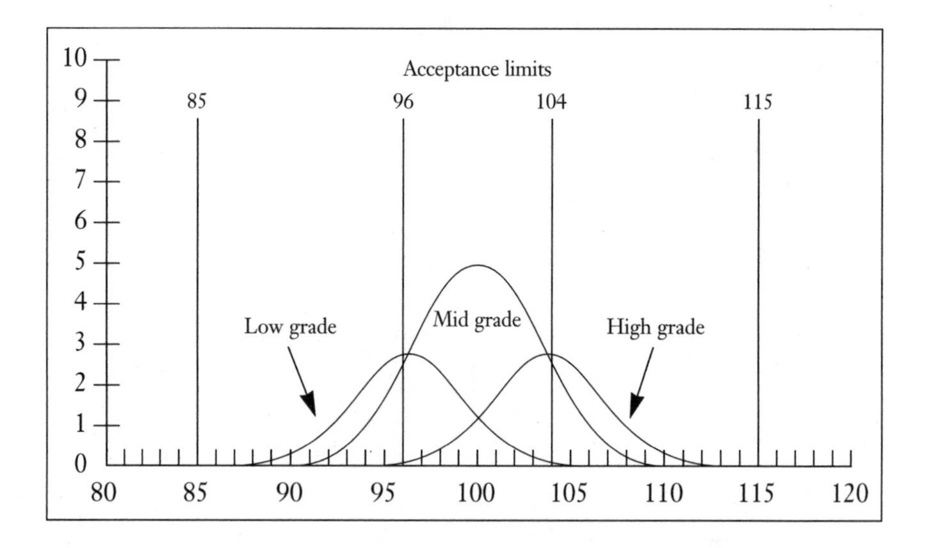

Figure 17.2. Product distribution by grading/sorting: Measurement variance = 60 percent of total variance.

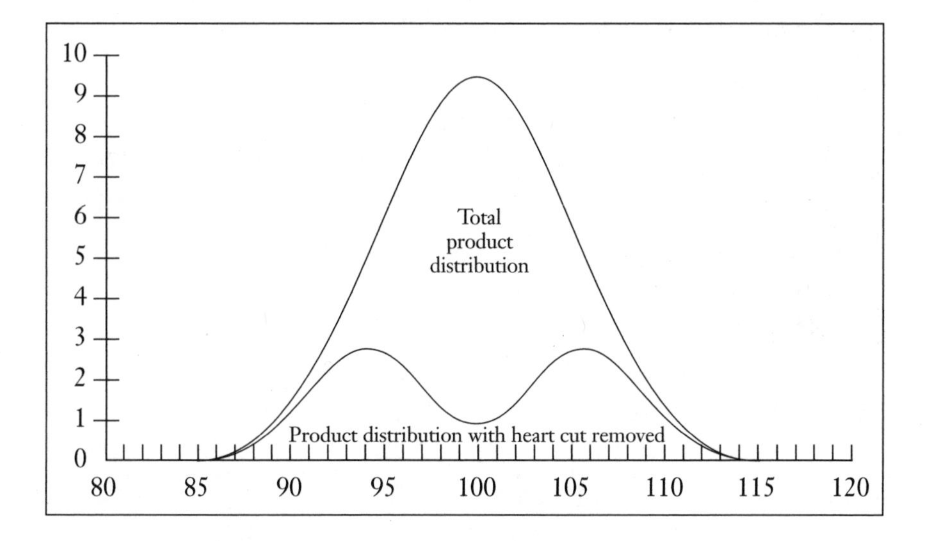

Figure 17.3. Product distribution with and without heart cut: Measurement variance = 10 percent of total variance.

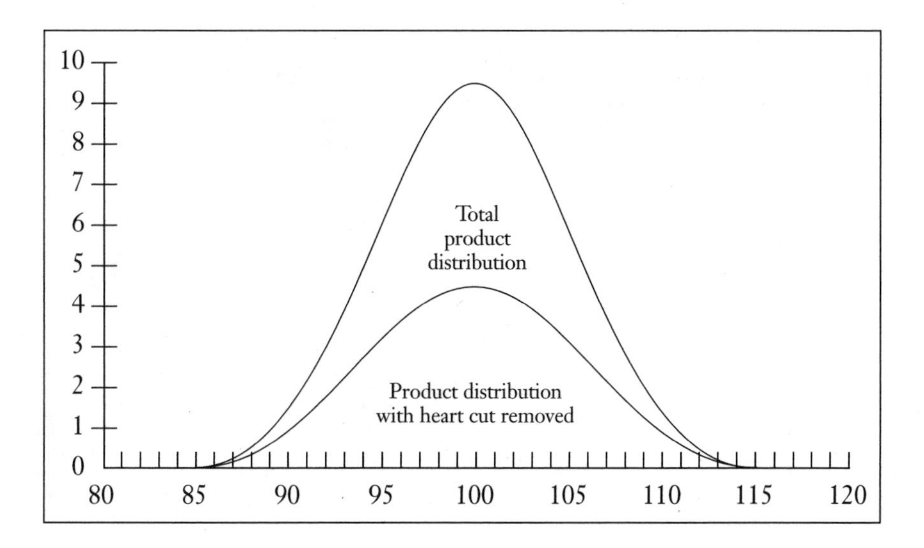

Figure 17.4. Product distribution with and without heart cut: Measurement variance = 60 percent of total variance.

Table 17.2. Distribution of sorted product.

	Category					
	– Low –		**– Mid –**		**– High –**	
Acceptance limits	85	96	96	104	104	115
Percent of product in each category						
10% measurement	18.5%		63.0%		18.5%	
60% measurement	27.5%		45.0%		27.5%	
Characteristic average within each category						
10% measurement	94.8		100.0		105.2	
60% measurement	97.1		100.0		102.9	
Limits including 99% of the product in each category						
10% measurement	87	101	93	107	99	113
60% measurement	88	106	91	109	94	112

If the effect of measurement variability changes, from changes either in the measurement procedure or in sample sizes, the average and spread of each of the categories will change.

CHAPTER 18

Resampling and Retesting

18.1 Resampling and Retesting

Resampling and/or retesting should not be used to replace sample results that indicate a nonconformity and, therefore, to make nonconforming product into conforming product. Resampling is

- Obtaining a completely new sample from the conforming product
- Appropriate to replace a sample that is known to be defective or invalid
- Appropriate to obtain additional information to determine the best disposition for nonconforming product

Retesting is

- Obtaining an additional test result on the same sample
- Usually less costly than resampling
- Appropriate to replace a test result that is known to be invalid

A test result outside the acceptance limits by itself is not a valid reason for resampling or retesting. Some examples where resampling or retesting is appropriate are

- Incorrect identification of sample
- Contaminated or damaged sample

- Faulty operation of sampling or testing equipment
- A test result grossly different from the expected value
- Sampling or testing technique known to be incorrect

When resampling or retesting is used to replace a known defective sample or test result, the original value should be retained and marked as replaced. The reason for taking the resample or retest should be recorded.

18.2 Double (or Multiple) Sampling

A double (or multiple) acceptance plan is an appropriate alternative to resampling and retesting. (See chapter 13.) Although a double acceptance plan may appear similar to a resampling procedure, the acceptance limits for a double-sampling procedure are different from those of a single-sampling plan because they have been calculated to accommodate the additional sampling risk.

A double-sampling plan may be designed two different ways. The acceptance limits are different for each procedure. One procedure or the other must be selected ahead of time.

- Use the average of the two or more sample results as the decision value. This option is appropriate when sampling and measurement variability is large.
- Use the second value alone for the decision. This option is appropriate when special causes or blunders are suspected.

CHAPTER 19

Waivers

19.1 Waivers

A waiver is the act of intentionally abandoning the requirement that the lot meets the specification. Also, a waiver is the document that details the reasons for this action. A waiver authorizes product to be delivered even though all specifications are not met. By affirming the waiver, both the producer and customer agree that the evaluation of the nonconforming product indicates that the product will perform essentially the same as conforming product and can be used as if it were conforming. The same procedure applies to both raw materials and final product. Good practice requires obtaining agreement from the customer before the product is delivered.

19.2 Product Released Under Waiver

Nonconforming product can be delivered to the customer under waiver only if the following criteria are met.

- Neither the nonconformity nor the reason for nonconformance is known adversely to affect the customer with respect to performance, reliability, or safety.
- The customer's end use would not be adversely affected by the nonconformity. This information should be obtained from the customer, if possible, and documented as a quality record.

- Agreement to deliver the product has been obtained from the customer.

19.3 Repeated Waivers

The repeated use of waivers to release product is a signal of a problem with either the producer's process or the customer's expectations. Two options are available for the producer.

- Re-assess the production process to identify and fix the problem areas.
- Consider renegotiating the specification limit with the customer. If the customer has been able to use the waver material without any problems, then wider specification limits may be justified.

Part V

Moving Ahead

CHAPTER 20

Supplier, Producer, and Customer Relations

20.1 Partnership and Trust

Success in business is best achieved by suppliers, producers, and customers working as partners. When the information necessary to carry out business flows freely and all parties trust this information, business practices become less complex. The result is lower costs for all parties.

Trust comes from

- Working together to understand each other's needs and capabilities in measurable terms
- Sharing test methods
- Making decisions based on facts
- Sharing information as generated (raw data)
- Agreeing on a common interpretation of the information
- Solving problems together
- Working together to meet future needs and for continuous improvement

20.2 Statistical information

Statistical interpretation of data is widely used to explain the relationships among variables. Such understanding can help achieve agreement on a common interpretation of information.

- Customers should beware of comparing the quality of different producers based upon a comparison of producer-provided product information. This is not a good practice unless the customer has determined that sampling, testing, and reporting practices are uniform among producers. Such practices can vary greatly and invalidate direct comparisons. The same caution applies to long-term evaluation of an individual producer. These measurement methods may also significantly change over time.

- Customers need to provide processability and finished product performance data so that the producer can determine the cause and effect between process parameters and product performance for the customer.

- Producers need to provide process performance data so that customers can understand the variability of the product and design their processes to accommodate for it.

Establishing specifications is a business practice that benefits from the free flow of information. Through information sharing, setting specification limits becomes a technical issue based on reaching the best match between the fitness-for-use requirements of the customer and the manufacturing capability of the producer.

In addition to the sharing of information on current products and process variability, working together to adopt new technology and business practices can bring the benefits of the new technologies and practices to all parties quickly and cost-effectively.

CHAPTER 21

Change Management

21.1 Generic Change Management

Managing changes that affect product quality is a very important aspect of a successful relationship. It is often the difference between a good producer-customer partnership and a great partnership.

There are always changes taking place within the producer's organization—manufacturing changes, raw material changes, organizational changes, business contract changes, and so on. The ability of the partners to pay attention to these details and communicate them to all parties involved keeps the partnership on track. The key element is that there be no surprises in product quality or end-use performance.

There are two aspects of change management related to the specification process: (1) specification document changes, and (2) technology changes. The terminology for these changes may vary from company to company, and a clear understanding of definitions by both the producer and customer is important.

It is recommended that the reader consult the ANSI/ISO/ASQC Q9000 series standards for requirements related to establishing a formal change management process. Another widely used reference for change management is OSHA CFR1910.119, also known as *Process Safety Management of Highly Hazardous Chemicals Standard, Title 29.*

21.2 Specification Document Changes

Specifications are dynamic documents that will change over time. It is important to have document control procedures that address initial development, routine changes, and rush (or emergency) changes. It is equally important to notify and obtain the customer's approval for all changes that could alter the final product or affect its use.

21.3 Technology Changes

Depending on specific company terminology, technology change may encompass a broader range of changes than specifications alone. Technology change is also dynamic and includes formulation, raw material, process, equipment, testing, sampling, shelf life, packaging, handling, labeling, manufacturing location, and so on. Most technology changes hold the potential for impacting specifications, other internal functions, and outside customers. Some of these changes may also be classified as design changes. Change management is applicable to all of these, and systems and procedures for handling them must exist.

21.4 Customer Notification and Approval

Customer notification of change prevents unwanted surprises and enhances the producer-customer partnership. Change notification requirements should not be viewed as inhibiting change. They are usually the result of technology improvements, and should be encouraged as continuous improvement.

Systems and procedures for carrying out the notification and approval process should exist, and records of customer notification and approval requirements should be maintained in support of this activity. The terms of notification and approval for changes should be established at the outset of business and revised as necessary. The need and extent of notification are generally dependent on the criticality of the material in the customer's application. Samples of material with the incorporated change must be made available for evaluation, and adequate inventories of material produced before the change must be maintained.

CHAPTER 22

Data Sharing

22.1 General

Data sharing is helpful for both the producer and customer. Product data helps the customer to better understand the cause-and-effect relationship between raw material properties and final product quality. Feedback data from the customer helps the producer improve process and product quality.

Data sharing may involve verbal, written, or electronic communication. All forms of communication require that each party know and understand the other party's definitions for the terms and the codes used.

22.2 Conformance Data Sharing

There are many different ways to describe conformance to specifications. These include the following:

- Certificates of analysis
- Certificates of compliance
- Quality summary reports
- Capability and performance data
- Statistical information

The parties involved must agree on the practices used and what the information means.

22.3 Certificates of Analysis

In the CPI, conformance to specification limits is often reported to customers through certificates of analysis (COA). A COA indicates that testing was performed and reports the results of those tests. A COA may also include the agreed-upon specifications. A COA is valuable for the customer, especially when used as a criterion for accepting the material. A COA must contain the information agreed upon between the customer and the producer.

The following elements are suggested on a certificate of analysis.

- Product identity (name and lot number)
- Test results
- Test method
- Specification limits
- Supplier order number
- Customer order number
- Shipment date
- Product origin (organization name and location)
- Product quantity
- Name and title of person certifying

The test results in a COA should be based on samples from the specified lot or actual material shipped. If they are not, the source of the test results should be specified. Agreement from the customer should be obtained and documented in cases where results are not based on samples from the actual lot of material shipped.

The contents of the COA should be verified, either by a qualified person or by software. Consideration should be given to the confidentiality of the information, usefulness of the information to the customer, and the ability to generate the required data.

22.4 Certificates of Compliance or Conformance

Certificates of compliance or conformance are documents certifying that the associated product or service has met the agreed-upon requirements.

These requirements may include specification limits, contractual obligations, government regulations, or conformance to a given quality standard. The producer must have systems in place to ensure conformance to requirements.

The COA may become a problem as producers strive to eliminate unneeded testing. As statistical process control improves the process, making it stable and predictable, the frequency of testing can be decreased using such techniques as skip-lot testing. Since the definition of a COA is testing conducted on a specific lot, there will be instances when no testing was conducted on a lot requiring a COA. The use of certificates of compliance or conformance will take on greater significance as these types of statements can be issued with skip-lot test plans. An alternative is to amend the definition of the COA to allow for the test data defining the production interval to be used.

The same considerations for content and approval for a COA apply to a certificate of compliance. Supporting test results, however, are not reported on a certificate of compliance.

22.5 Quality Summary Reports and Other Statistical Reports

The purpose of a quality summary report is to provide data to support decisions. Any form of data can be referenced or included in a quality report. Often, control charts or histograms showing the variability of the process or product characteristics are included.

Statistical analysis of data provides additional understanding of the process that produced the product and of the true quality or nature of the product. Part of the process of establishing trust for the statistical analysis is to clearly define the statistics being requested and how they will be used. The following points must be resolved.

- Whether process or product information is to be supplied.
- For process information, the data sharing agreement should cover protection of proprietary information.
- For product information, critical properties and the sampling plan must be described, preferably in the product specification.

- If control charts are requested, the type of control charts, subgroup size, and formulas for calculations should be supplied.
- The criteria for out-of-control observations should be supplied.
- Statistical data for the test methods are often helpful and may provide insight for the interpretation of the other data supplied.

22.6 Electronic Data Sharing

Electronic data transfer introduces complex verification issues, especially for certificates of analysis and certificates of compliance. Verifications include those of

- The sender
- Communication capability (format, compatible hardware and software, and so on)
- Data entry
- Software

22.7 Software Verification

Data are frequently gathered, stored, and transmitted electronically. Conformance decisions are now often made by software, and the results are stored and transmitted electronically. Similar to mechanical systems that provide data require calibration, software systems that store, manipulate, and transmit data also need verification. Software verification has the following elements:

• *Development verification*—A system for the development, acceptance, and implementation of software must be established. The system must address how software is to be tested, authorizations and approvals that are required, and the method of acceptance designated.

• *Test*—Criteria for demonstrating that the software is adequate for its intended use must be developed, and the software tested against these criteria.

• *Documentation*—There should be sufficient documentation to provide the basis for software maintenance.

- *Library controls*—Procedures must ensure that different software versions are identified; that no unauthorized changes are made; that all approved changes are properly incorporated; and that the software in use is the correct, approved version.
- *Responsibilities*—Responsibilities for authorizations and approvals of the above elements and for developing and carrying out verification tests must be established. Authorization for maintenance and changes to software must also be established.
- *Audits*—Audits of the software must be routinely performed, and audit records retained.

CHAPTER 23

Communication Methods

23.1 General

Today the number of communication methods has expanded beyond the traditional paper document and spoken word to include a group of methods labeled *electronic commerce.* Each form of communication provides specific benefits and limitations. The essential requirements are that communication be explicit and that there be adequate permanent records of the communication on paper or electronic media. These requirements must be met if the benefits of the communication method are to be achieved and the communication is to be effective.

All types of communication require that each party understand the other's definitions for the words and codes used. Language considerations are critical. It is essential that all new terms be defined when first used, and that the partner in the communication confirm that the definitions are understood.

23.2 Signatures

In business a signature stands for the concept of approval and/or authorization. A person's handwritten signature is considered to be a unique, nonforgeable symbol of authorization and approval. Paper documents remain the only means of transmitting an original handwritten signature. Authorization and approval can be communicated by other symbols and means, provided these ensure traceability to the person responsible and ensure integrity of the action.

23.3 Spoken Word

The spoken word—which may also be transmitted electronically by telephone, voice mail, and so on—provides important insight that is difficult to transmit otherwise such as the benefit of the tone, rate, and emphasis with which the words are delivered. The spoken word requires skill to be effective.

The unrecorded spoken word does not provide a permanent communication record. Agreements must be permanently recorded on paper or electronic media.

23.4 Written Communication

Written communication, which may also be transmitted electronically by fax (facsimile transmission), electronic mail, and so on, is the traditional form of business communication. The best-written communications are simple and direct and use well-designed formats to transmit data and other information.

23.5 Electronic Data Interchange

Electronic data interchange is defined as direct communication of data through electronic communication networks and other remote connections, generally under the influence of software that is also capable of storing and manipulating the data. It is different from the electronic transmission of spoken and written communication through voice mail, electronic mail, and fax that is widely used today. The safeguards described in section 22.6 must be employed.

Part VI

Case Study

CHAPTER 24

A Case Study

24.1 Introduction

The purpose of this case study is to illustrate the concepts and techniques discussed in the previous chapters. It is an example of how these concepts and techniques can be integrated into a meaningful and effective specification system. The case study is of a fictitious company that has features found in many process industries.

XYZ Company produces several types of snacks—potato chips, crackers, and so on. XYZ is respected throughout the trade as a high-quality producer. ABC Company, a nationwide food distributor, has contracted with XYZ to produce a new potato snack called "ThanQs." ABC is the customer.

ThanQs will be produced by grinding potatoes, mixing them with other ingredients, and extruding the snacks in the shape of the letter *Q*. The Qs are then baked and packed in plastic bags. XYZ needs to prepare specifications to ensure consistent high quality. XYZ is developing raw material specifications and process specifications internally. XYZ and ABC are jointly developing the product specifications.

24.2 The Process

A schematic diagram of the process is shown in Figure 24.1. The manufacturing process, as shown, has been simplified to illustrate the principles and procedures associated with specifications.

- *Batch processes*—Batch processes are common in the CPI. Some batch processes are primarily blending operations, and others involve multiple reactions and complex chemical treatments.

- *Extrusion processes*—Extrusion processes are common in the CPI (fibers, films, piping, and so on). Pretreatment of material before extrusion often consists of cooking or melting.

- *Bagging processes*—Bagging, packaging, and sealing operations are common in the CPI. Liquids (paints and solvents) are packed in small cans or drums. Solids (powders, granules, and fibers) are packed in bags, bales, cartons, or boxes.

24.2.1 The Batch Makeup

The first part of the process is a batch process in which potatoes are ground and then mixed with seasonings, binders, preservatives, and other ingredients. The pulp then goes through a solids-control unit where it is mixed with water to give it the flow capability to move easily from the batch tanks to the extruder step. The batch tanks are filled and used one at a time. The product in the batch tanks tends to discolor, and it makes unacceptable chips if left in the tanks for more than two hours.

Raw material specifications are needed for the

- Potatoes
- Additive A
- Additive B

Process specifications are needed for the

- Batch makeup
  - ♦ Potatoes (kilograms)
  - ♦ Additive A (number of 500 gram packets)
  - ♦ Additive B (kilograms)

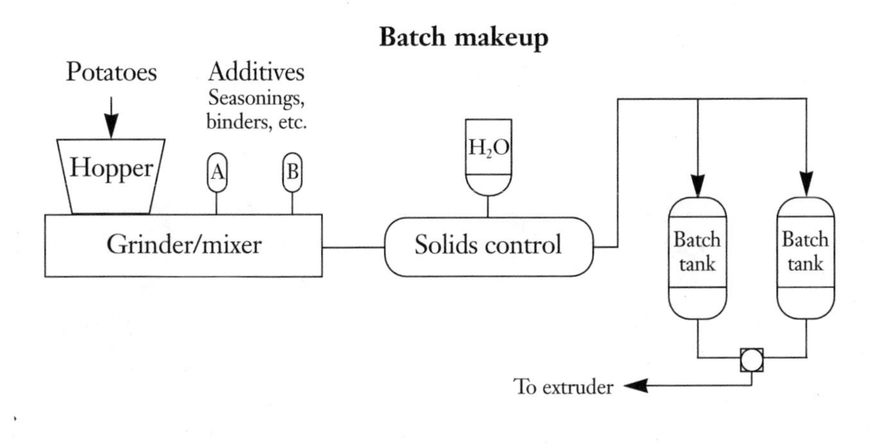

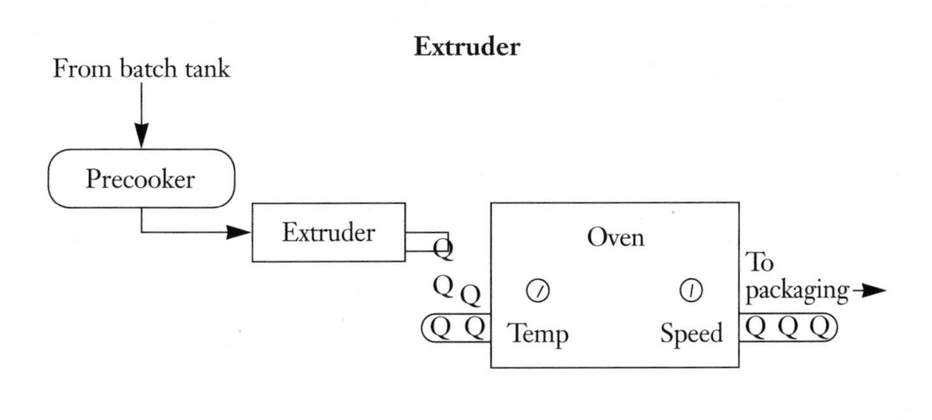

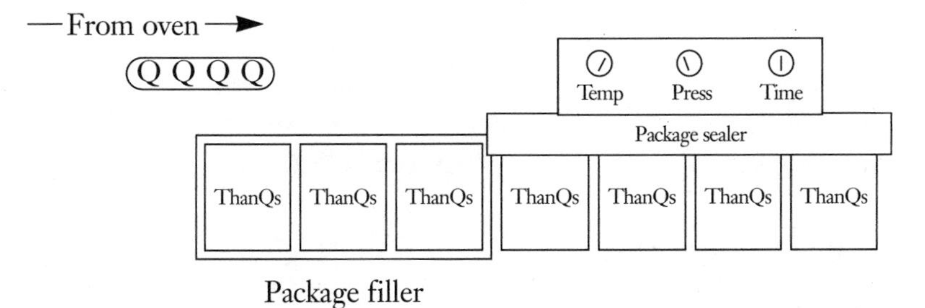

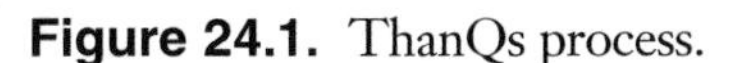

Figure 24.1. ThanQs process.

- Maximum batch hold time (minutes)
- Solids (percent)

24.2.2 The Extruder

The pulp is pumped from one of the batch tanks, then cooked partially to remove water and to provide the proper consistency for extrusion into Qs. The extruded Qs are deposited on a moving belt that passes through an oven to finish cooking and generate the desired crispness. The Qs then continue to the bagging operation.

Process specifications are needed for

- Precooker
  - Charge size (kilograms)
  - Steam pressure (kilograms/centimeter^2)
  - Time (minutes)
  - Solids to extruder (percent)
- Extruder
  - Temperature (°C)
  - Pressure (kilograms/centimeter^2)
- Oven
  - Belt speed (meters/minute)
  - Temperature (°C)

Product specifications are needed for

- Crispness
  - Breaking strength (grams)
- Appearance
  - Color, breakage (percent defective)

24.2.3 The Bagging Operation

The completed Qs arrive on the belt from the extruder and oven. Bags are filled with Qs and then passed through a sealer.

Raw material specifications are needed for the

- Bags
  - Thickness (mils)
  - Puncture resistance (grams)

Process specifications are needed for the

- Bag filler
  - Speed (meters/minute)
- Bag sealer
  - Platen temperature (°C)
  - Platen pressure (kilograms/centimeter^2)
  - Sealing time (seconds)

Product specifications are needed for the

- Bags
  - Weight (grams)
- Seals
  - Strength (grams)

24.3 The Raw Material Specification

Figure 24.2 is a raw material specification. It includes the requirements for potatoes, additives, and bags.

• Previous work has indicated that the best potatoes for both processing and flavor are Kennebec or equivalent. The top quality grade (U.S. Extra No. 1) is required.

• Flavors, Inc. has developed a wide range of premixed and prepackaged spices, flavors, and other ingredients. Ingredient #2200A was selected for additive A as a result of a joint test program between Flavors, Inc., XYZ, and ABC. Additive B is a dry mixture of salt and citric acid $C_3H_4OH[COOH]_3$, which is prepared as a solution as required. The citric acid adds flavor and helps to minimize discoloration.

• The bagging material is made from recycled plastic containers. PSTX is the only supplier making a material that is approved for food

XYZ Company—Sargentville Plant

Raw material specification

Specification number M 132 Date: 2/9/96

Revision number 0 Author: Spud Baker

Approvals: Jennifer Sanders

Product: ThanQs

Item	Description	Specification limits Low	Target	High	Measurement units	Test method
1	Potatoes	Kennebec or equivalent—U.S. Extra No.1				
2	Additive A	Flavors, Inc. #2200A—500 gram packet				
3	Additive B—NaCl,	46.0	47.0	48.0	percent	TM135
	$C_3H_4OH[COOH]_3$	41.0	42.0	43.0	percent	TM136
4	ThanQ bags (PSTX Corp. only approved supplier)					
	Thickness	6.5	7.0	7.5	mils	TM005
	Puncture resistance	500	800	1600	grams	TM220

Comments:

a. Additive A is a proprietary mixture of Flavors, Inc.

b. PSTX is the only approved supplier for recycled plastic products used for bagging food products.

Figure 24.2. Raw material specification.

bagging. PSTX has agreed to meet the thickness specification limits of 7.0 ± 0.5 mils by TM005 and a puncture resistance target of 800 grams by TM220 with a minimum specification limit of 500 grams. The upper specification limit for puncture resistance has been set at 1600 grams. While very high puncture resistance is not a performance problem, it may indicate a change in untested properties. PSTX will also supply control chart information on a monthly basis and performance and capability information as requested by XYZ.

24.3.1 The Purchase Specification

Purchase specifications are prepared to inform an individual supplier of the requirements for purchased products. Figure 24.3 is the purchase specification for additive B.

- Items 1 and 2 are identical to the raw material specification.
- Item 3 specifies the package weight.
- The comments list labeling, packaging, and other requirements.

24.4 The Process Specification

24.4.1 Batch Makeup

Figure 24.4 is the process specification for batch makeup.

- The required mixture of items 1, 2, and 3 was developed during the joint test with ABC, XYZ, and Flavors, Inc.

- A solids-content level of 25 percent is required to achieve the proper consistency for transport from the batch tanks to the precooker. Samples for solids content are taken at the start of each batch and 10 minutes into the batch.

- After two hours in the batch tank, the pulp tends to discolor. The comment section includes the instruction not to use the batch if the maximum hold time of two hours is exceeded.

XYZ Company—Sargentville Plant

Purchase specification

Specification number PS 132
Revision number 0

Date: 2/12/96
Author: Spud Baker
Approvals: Roy Hancock

Supplier: Eggemoggin Chemical Supply
Location: Brooksville Plant

Supplier acceptance:
Name: Robert Burton
Title: QA Manager

Raw material name: Additive B
Trade name: Seasoning mix 1515 (food grade)

		Specification limits			Measurement	Test
Item	Description	Low	Target	High	units	method
1	NaCl	46.0	47.0	48.0	percent	TM135
2	$C_3H_4OH[COOH]_3$	41.0	42.0	43.0	percent	TM136
3	Net weight	49.5	50.0	50.5	pounds	Scale wt.

Comments:

a. Label bags "XYZ—Additive B"

b. Stack bags eight high on pallets. Shrink wrap pallets.

c. Include a certificate of analysis with each shipment.

d. Notify XYZ Company and obtain approval for any significant process changes.

e. XYZ Company reserves the right to audit the supplier's quality system and/or request evidence of statistical process control.

Figure 24.3. Purchased products specification for additive B.

XYZ Company—Sargentville Plant
Process specification

Specification number B 361 | Date: 2/14/96
Revision number 4 | Author: Spud Baker

Product: ThanQs
Mfg. area Batch makeup

Approvals: Charlotte Austin

Item	Description	Specification limits Low	Target	High	Measurement units	Test method
1	Potatoes	240	250	260	kg	Scale S5
2	Additive A 500 gram packet	1	1	1	packet	as added
3	Additive B	3.95	4.00	4.05	kg	Scale S7
	Process parameter					
4	Solid contents to batch tank	22	25	28	% solids	TM106
5	Hold time in batch tank	30	60	120	minutes	Timer B3

Comments:

a. Take samples for solids at the beginning of a batch and after 10 minutes.

b. Material must not be used for ThanQs if left in batch tank more than 2 hours.

Figure 24.4. Process specification for batch makeup.

24.4.2 Extruder

Figure 24.5 is a process specification for the extruder area.

- The target values for the precooker, extruder, and oven parameters were developed in an optimization experiment by research and development (R&D) using the principles of statistical design of experiments. (See Box, Hunter, and Hunter 1978).

- The precooker is used to remove water and to provide some cooking to consolidate the pulp into the consistency necessary for the extruder. The precooker solids content and the ThanQ crispness are measured on samples taken from each batch and are monitored by real-time control charts. Precooker time is used to control solids content.

- The extruder conditions determine the shape and the texture of the Qs.

- Belt speed determines the cooking time and the production rate.

- Oven temperature is used to control crispness. A wide range of oven temperatures may be required. The specification shows the limits for both the set point and the limits around the set point.

24.4.3 Bagging

Figure 24.6 is a process specification for the bagging area.

- As the bags pass through the sealer, a heated platen is applied under pressure. The sealing time is the length of time that the platen is in contact with the bag.

- The important process parameters in the bagging operation are the sealer settings and the bag-filler speed. Seal strength is controlled by sealing time.

24.5 The Product Specification

Figure 24.7 is the product specification. It shows limits for all of the product characteristics. These have been negotiated with the customer, ABC Company. ABC requested that the upper specification limit for percent defective ThanQs be set at 5 percent. XYZ company agreed and will use a p chart with a target of 1 percent to monitor the process.

XYZ Company—Sargentville Plant

Process specification

Specification number E 819 — Date: 2/29/96

Revision number 2 — Author: Spud Baker

Product: ThanQs

Mfg. area Extruder

Approvals: Winfield Lowell

Joanna Lee

Item	Process parameter	Specification limits Low	Target	High	Measurement units	Test method
1	Precooker charge size	125	135	145	kg	Scale S9
2	Precooker steam pressure	7.25	7.75	8.25	kg/cm^2	Meter PR101
3	Precooker time	12	15	18	min	Timer E6
4	Solids to extruder	60	64	68	%	TM106
5	Extruder temperature	88	90	92	°C	Meter TR51
6	Extruder pressure	2.40	2.50	2.60	kg/cm^2	Meter PR102
7	Belt speed	14	15	16	m/min	Meter SP12
8	Oven temperature					
	Set point	200	225	250	°C	Meter TR21
	At set point	–5	0	+5	°C	Meter TR21

Comments:

a. Precooker solids content is controlled by precooker time.

b. Crispness is controlled by oven temperature.

c. Notify quality engineer if control charts indicate adjustments outside the specification limits.

Figure 24.5. Process specification for the extruder.

XYZ Company—Sargentville Plant
Process specification

Specification number B 744 | Date: 2/26/96
Revision number 1 | Author: Spud Baker

Product: ThanQs
Mfg. area Bagging

Approvals: Yvonne Bradford

Item	Process parameter	Specification limits Low	Target	High	Measurement units	Test method
1	Sealer platen temperature	145	150	155	°C	Meter TM371
2	Sealer platen pressure	3.00	3.15	3.30	kg/cm^2	Meter PR119
3	Sealer sealing time	2.0	3.0	4.0	sec	Timer B9
4	Bag-filler speed	50	70	90	m/min	Meter S23

Comments:

a. Seal strength is controlled by sealing time.

Figure 24.6. Process specification for bagging.

24.5.1 Quality Measures—Defective Qs

The finished ThanQs are inspected at the oven's exit. A sample of 200 ThanQs is checked for defects during each production hour. One hour's production has been chosen as a lot. TM865 describes the types of defects to be counted. A percent-defective control chart (*p* chart) is used to detect changes in the level of defectives. With a target of 1.0 percent defective, the upper control limit is 3.1 percent allowing some room to

XYZ Company—Sargentville Plant
Product specification

Specification number FP 644 Date: 2/28/96
Revision number 2 Author: Eaton Wright

Product: ThanQs

Approvals: June Vaughn

Item	Product characteristic	Product unit	Specification limits Low	Target	High	Measurement units	Test method
1	Defective ThanQs	bag	0	1	5	%	TM865
2	Crispness	ThanQ	125	140	155	grams	TM848
Bagging							
3	Net weight	bag	290	300	310	grams	TM866
4	Bag seal strength	bag	1300	1500	1700	grams	TM812

Test frequency		Sample size	Frequency
1	Defective ThanQs	200 Qs	1/hour
2	Crispness	10 Qs	1/hour
3	Net weight	10 bags	1/hour
4	Bag seal strength	4 bags	1/hour

Comments:

a. Net weight is controlled by bag-filler speed.

b. The lot for acceptance is 1 hour of production.

c. Control limits for the p chart for defective ThanQs are

Upper control limit	3.1%
Target	1.0%
Lower control limit	0.0%

Figure 24.7. Product specification.

detect an increase and take corrective action before producing ThanQs out of specification limits.

The quality engineer chose a single sample acceptance sampling plan as an appropriate method for identifying lots that should be rejected from shipment because of a large number of defective Qs. The plan used is taken from ANSI/ASQC Z1.4-1993. The code letter L plan with AQL = 1.0 gives a probability of acceptance of 10 percent at 4.64 percent defective. (The OC curve is shown on page 51 of the standard.)

24.5.2 Quality Measures—Crispness

Crispness is measured by TM848, a proprietary method that involves breaking the ThanQs. The results are entered into a variables control chart for crispness. The target of 140 was determined from the R&D experiments. The specification limits have been set to represent the process performance index of 1.33 because the customer has not requested limits (see chapter 14).

The process variability study gave the following components of variance.

Source	**Variance**		
Batch-to-batch	3.4		
ThanQs within batch	6.3	Total product	9.7
Long-term measurement	0.8		
Short-term measurement	3.1	Total measurement	3.9
Grand total	13.6		13.6

The ±4σ limits have been estimated using the square root of the grand total variance. The limits are

$$\pm 4s_{total} = \pm 4\sqrt{13.6} = \pm(4)(3.69) = \pm 14.8 \text{ (Round to 15)}$$

The specification limits for individual Qs are

$$\text{Lower specification limit} = 140 - 15 = 125$$

$$\text{Upper specification limit} = 140 + 15 = 155$$

Statistical process control will be used to keep the crispness on target.

24.5.3 Bagging

Seal strength is extremely important. If the seal strength is too low, bags can easily open in transit or handling before use. If the seal strength is too high, it becomes difficult to open the bag by hand. The final customer's needs are related to the ease of opening the bag. A seal strength target of 1500 grams has been selected to provide a balance between the pre-opening and the ease-of-opening requirements. Specification limits of 1300 grams to 1700 grams have been set as the amount of variation the customer can tolerate.

The specification for the net weight of product in the bag is important to give the customer fair value and to prevent product damage due to overpacking. ABC has requested a 300-gram package for the intended market. The specification limits of 290 grams to 310 grams have been recommended by the manufacturer of the packaging equipment. Both ABC and XYZ agree that these limits are appropriate.

24.6 Lot Acceptance Limits

Although the specification limits apply to individual units, Qs or bags, the product is accepted as one-hour lots including about 2000 bags. A sample of several units is tested or examined and the average used for the decision to accept or reject the lot. The acceptance limits are different from the specification limits to allow for variability in sampling and measurement. The OC curve calculation is used to determine the degree of offset from the specification limits under the assumption of suitable risks.

The lot acceptance limits for bag seal strength are 1390 grams to 1610 grams for a sample size of four bags. The calculation for these limits is included in Exhibit A. The variance components have been estimated from a sampling study using the test plan described in chapter 10. The data are shown in Table 24A.2. The variance components are shown in Table 24A.5. Trial calculations have been carried out for sample sizes of one and four. The OC curves are shown in Figure 24A.2.

The sample size of four bags per lot has been chosen because it provides a 57 percent wider operating interval for the producer while maintaining the customer's risk at no more than 2.5 percent. XYZ has determined that the wider operation interval is worth the added cost of testing four bags per lot versus one bag. Similar calculations can be carried out for crispness and net weight.

EXHIBIT A

The OC Curve for Seal Strength

24A.1 The Data

The sampling plan used to collect the data for the variance components is the balanced plan discussed in chapter 10. The objective was to determine the variance components, develop a two-sided acceptance sampling plan, and plot its OC curve. The following plan shows how the process was sampled and the seal strength was measured.

Each day for 40 days,

- Select a sealed bag from the end of the bagger.
- One hour later, select another bag from the end of the bagger.
- Cut four, 1-inch samples from the seal area of each bag. Identify two samples as test-time-1 and the other two as test-time-2.
- Measure the seal strength for the two test-time-1 samples.
- At a later time (day or shift), measure the seal strength for the two test-time-2 samples.

Table 24A.1 identifies the eight test samples taken on each day. These samples appear as a single row in Table 24A.2.

Table 24A.1. Daily sample schedule.

Sample result	Bag	Test time (LT)	Measurement (ST)
1	1	1	1
2	1	1	2
3	1	2	1
4	1	2	2
5	2	1	1
6	2	1	2
7	2	2	1
8	2	2	1

24A.2 The Analysis of Variance

A nested model was used to obtain the analysis of variance (ANOVA) table and the variance components. The ANOVA table is shown in Table 24A.3.

$$\text{Seal strength} = \mu + \text{Lot}_i + \text{Bag}_{j(i)} + \text{LT}_{k(ij)} + \text{ST}_{l(ijk)}$$

The variance component structure of the expected means squares for this model are given in Table 24A.4. The expected means squares were solved to give the variance components. They are shown in Table 24A.5.

The variance components can be grouped by product and measurement in Table 24A.6. Note that the standard deviations are the square root of the variances. They are not additive.

24A.3 Process Performance

Three process performance indexes have all been computed using the range of the specification limits (1700 – 1300 = 400) and a standard deviation from the variance components.

Table 24A.2. Data for analysis of variance.

	Bag 1				Bag 2			
	Time 1		Time 2		Time 1		Time 2	
Lot 1	1446	1455	1470	1501	1511	1486	1509	1514
2	1455	1462	1475	1491	1418	1439	1445	1435
3	1354	1400	1411	1384	1432	1443	1432	1438
4	1610	1632	1627	1624	1630	1669	1624	1658
5	1363	1418	1413	1421	1331	1374	1308	1361
6	1503	1475	1453	1454	1518	1548	1490	1456
7	1527	1538	1517	1542	1482	1495	1501	1473
8	1502	1574	1517	1530	1567	1581	1576	1554
9	1440	1428	1426	1456	1490	1466	1466	1476
10	1431	1410	1431	1458	1396	1378	1441	1418
11	1697	1683	1620	1630	1668	1626	1652	1629
12	1652	1712	1674	1686	1636	1622	1648	1617
13	1592	1593	1585	1585	1524	1509	1570	1524
14	1346	1339	1399	1339	1283	1285	1323	1305
15	1452	1435	1481	1444	1561	1512	1482	1536
16	1564	1537	1563	1573	1579	1523	1582	1571
17	1463	1426	1458	1454	1410	1395	1432	1460
18	1467	1440	1454	1513	1503	1489	1512	1513
19	1597	1585	1601	1586	1563	1587	1577	1572
20	1501	1534	1489	1482	1386	1449	1412	1385
21	1556	1616	1621	1600	1601	1562	1604	1562
22	1496	1513	1520	1531	1496	1519	1497	1465
23	1469	1445	1481	1476	1468	1433	1490	1494
24	1499	1480	1466	1476	1499	1486	1476	1481
25	1610	1622	1616	1583	1670	1695	1703	1690
26	1606	1588	1634	1606	1660	1652	1671	1659
27	1598	1558	1612	1616	1558	1595	1540	1590
28	1404	1424	1417	1413	1428	1441	1447	1462
29	1404	1429	1439	1454	1399	1381	1409	1431
30	1297	1323	1302	1333	1346	1367	1304	1321
31	1535	1519	1487	1498	1421	1483	1427	1464
32	1563	1524	1570	1570	1498	1551	1558	1525
33	1502	1507	1520	1475	1509	1492	1490	1451
34	1568	1555	1594	1553	1522	1507	1496	1518
35	1483	1543	1551	1515	1516	1565	1531	1554
36	1651	1632	1670	1672	1580	1576	1591	1616
37	1562	1629	1568	1564	1653	1635	1645	1623
38	1582	1538	1601	1579	1548	1638	1594	1611
39	1413	1428	1427	1449	1482	1469	1462	1497
40	1535	1519	1572	1560	1516	1579	1569	1589

Table 24A.3. ANOVA table for bag seal strength.

Source	Degrees of freedom		Sum of squares	Mean square	*F* ratio
Lots	L – 1	39	345312	8854	2.97
Bags	(B – 1)L	40	119238	2981	4.46
Long-term measurement	(T – 1)LB	80	53402	668	1.30
Short-term measurement	(R – 1)LBT	160	82224	514	
Total	LBTR – 1	319	600176		

Table 24A.4. Expected mean squares for ANOVA for bag seal strength.

Source	Mean square	Expected mean squares
Lot-to-lot	MS lots	$\sigma^2_{ST} + R\sigma^2_{LT} + RT\sigma^2_{BAG} + RTB\sigma^2_{LOT}$
Bags within lot	MS bags	$\sigma^2_{ST} + R\sigma^2_{LT} + RT\sigma^2_{BAG}$
Long-term meas	MS long term	$\sigma^2_{ST} + R\sigma^2_{LT}$
Short-term meas	MS short term	σ^2_{ST}

Where L = Number of lots sampled = 40
B = Number of bags sampled per lot = 2
T = Number of different times each bag was measured = 2
R = Number of replicates made at each measurement time = 2

Table 24A.5. Variance components and standard deviations for bag seal strength.

Variance component	Variance (s^2)	Standard deviation(s)
Lot-to-lot	734	27.1
Bags within lot	578	24.0
Long-term measurement	77	8.8
Short-term measurement	514	22.7
Total	1903	43.6

Table 24A.6. Grouped variances and standard deviations for bag seal strength.

Variance component	Variance (s^2)	Standard deviation(s)
Total product	1312	36.2
Total measurement	591	24.3
Grand total	1903	43.6

- The performance index based on the total variability is

$$P_p = 400 / [(6)(43.6)] = 1.53$$

- The performance index for the measurement process is

$$P_t = 400 / [(6)(24.3)] = 2.74$$

- The performance index for the product variation only is

$$P_{prod} = 400 / [(6)(36.2)] = 1.84$$

The performance indexes indicate that the process is capable of meeting the specification limits. The P_p of 1.53 is greater than the level of 1.33 that is often used to define acceptable capability. The largest variance component is the lot-to-lot variability. Continual improvement effort focused on lot-to-lot variability is appropriate. The variability of bags within lot is also large. It emphasizes the need to keep the process on target and to reduce within-lot variability. The use of control charts will aid in keeping the process on target. The measurement variability is a potential problem with a P_t of 2.74. Measurement variance is 31 percent of the total variance. Continual improvement effort on the measurement process is also appropriate. (Herman 1989, 670–675.)

24A.4 The OC Curve

The XYZ and ABC companies have set the requirement for an acceptable lot as one with no more than 2.5 percent of bags out of specification limits for seal strength. Therefore, the minimum acceptable lot average has been set at $2s_{BAG}$ (bags within lot) higher than the lower specification limit. The maximum acceptable lot average is set at $2s_{BAG}$ lower than the upper specification limit (Figure 24A.1).

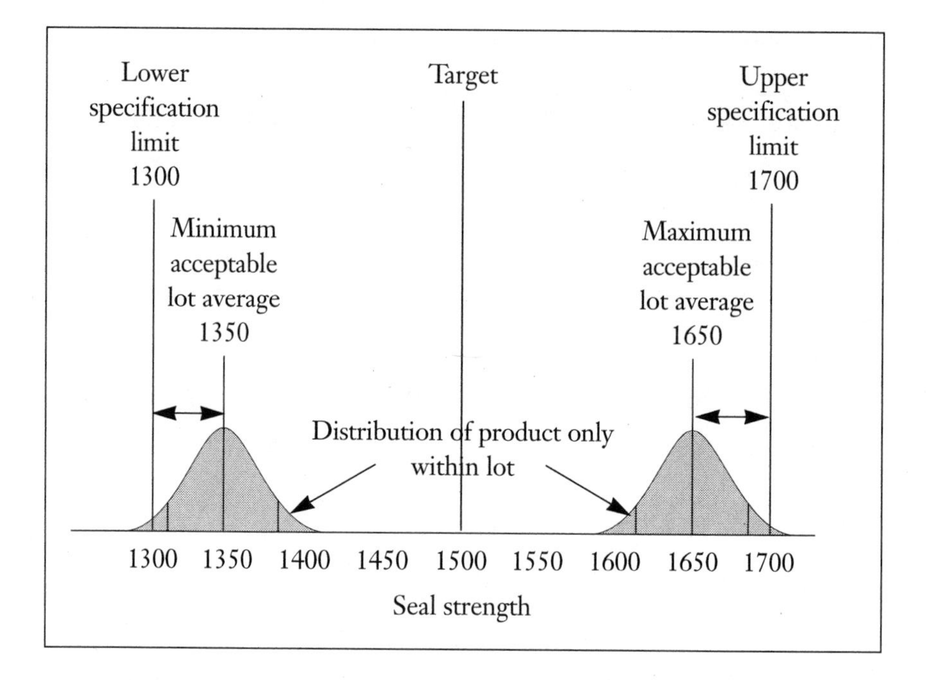

Figure 24A.1. Relationship between specification limits and acceptable lot averages.

If the true process average is at the minimum or maximum acceptable lot average, approximately 97.5 percent of bags with seal strength will be within the specification limits.

$$2s_{BAG} = (2)(24.0) = 48.0 \text{ (Round to 50)}$$

Upper specification limit = 1700

Maximum acceptable lot average limit = 1700 − 50 = 1650

Lower specification limit = 1300

Minimum acceptable lot average limit = 1300 + 50 = 1350

These are the maximum and minimum limits for the lot average.

24A.4.1 The Acceptance Sampling Plan

Two sampling and measurement plans were evaluated. OC curves were used to define and evaluate the acceptance plan. A lot is one hour of production (approximately 2000 bags).

Acceptance plan 1:

- Select four bags throughout lot.
- Send the selected bags to the test station for seal strength measurement. Each bag is measured using a 1-inch sample from the seal.
- Average the seal strength results.
- Accept or reject each lot by comparing the reported lot average with the lot average acceptance limits.

Acceptance plan 2:

- Same as plan 1 except sample and measure only one bag per lot and different acceptance limits.

24A.4.2 Variability of Lot Average Estimate

The standard deviation used in the computations for the OC curve is that for the variability of the reported lot average.

$$s^2_{avg} = \frac{s^2_{BAG}}{n_{BAG}} + \frac{s^2_{LT}}{n_{LT}} + \frac{s^2_{ST}}{n_{ST}}$$

where

n_{BAG} = Number of bags represented in the reported lot average
n_{LT} = Number of different test times represented in the reported lot average
n_{ST} = Number of different test results represented in the reported lot average

The variance and the standard deviation for the two proposed sample plans are given in Table 24A.7.

Table 24A.7. Variances and standard deviations sample plans for bag seal strength.

Sample plan	Variance	Standard deviation
Average 4 bags/lot	(578/4 + 77/1 + 514/4) = 350	18.7
One bag/lot	(578/1 + 77/1 + 514/1) = 1169	34.2

24A.4.3 Developing the OC Curve (4 bags/lot)

The desired P_{ac} at the lower lot average limit = 0.01

z = –2.33 (from the normal curve tables for 0.01 probability)

The standard deviation of lot average (sample size = 4) = 18.7

Lower acceptance limit (LAL) = Lower lot average limit – $z\sigma_{AVG}$
= 1350 – (–2.33)(18.7)
= 1350 + 43.6
= 1394.6 (Round to 1390.)

The desired P_{ac} at the upper lot average limit = 0.01

z = –2.33 (from the normal curve tables for 0.01 probability)

Upper acceptance limit (UAL) = Upper lot average limit + z_{AVG}
= 1650 + (–2.33)(18.7)
= 1650 – 43.6
= 1606.4 (Round to 1610.)

The probability of acceptance P_{ac} for both lower and upper acceptance limits has been computed for a range of true lot averages.

$$z_{lower} = (\text{Lot average} - \text{LAL}) / \sigma_{AVG}$$

$$z_{upper} = (\text{UAL} - \text{Lot average}) / \sigma_{AVG}$$

$$P_{re}(l) = 1 - P_{ac}(l)$$

$$P_{re}(u) = 1 - P_{ac}(u)$$

$$P_{ac} = 1 - P_{re}(l) - P_{re}(u)$$

Table 24A.8 shows the computations for several points on the OC curve. The values for the specification limits (LSL, USL), the minimum and maximum acceptable lot averages, the lower and upper acceptance limits (LAL, UAL), and the target are shown in bold type.

Table 24A.8. Computations for OC curve—4 bags/lot.

	Lot avg	z_{lower}	z_{upper}	$P_{ac}(l)$	$P_{ac}(u)$	$P_{re}(l)$	$P_{re}(u)$	P_{ac}
LSL	**1300**	**–4.813**	**16.578**	**0.0000**	**1.0000**	**1.0000**	**0.0000**	**0.0000**
	1325	–3.476	15.241	0.0003	1.0000	0.9997	0.0000	0.0003
Min lot avg.	**1350**	**–2.139**	**13.904**	**0.0163**	**1.0000**	**0.9837**	**0.0000**	**0.0163**
	1375	–0.802	12.567	0.2113	1.0000	0.7887	0.0000	0.2113
LAL	**1390**	**0.000**	**11.765**	**0.5000**	**1.0000**	**0.5000**	**0.0000**	**0.5000**
	1400	0.535	11.230	0.7037	1.0000	0.2963	0.0000	0.7037
	1425	1.872	9.893	0.9694	1.0000	0.0306	0.0000	0.9694
	1450	3.209	8.556	0.9993	1.0000	0.0007	0.0000	0.9993
	1475	4.545	7.219	1.0000	1.0000	0.0000	0.0000	1.0000
Target	**1500**	**5.882**	**5.882**	**1.0000**	**1.0000**	**0.0000**	**0.0000**	**1.0000**
	1525	7.219	4.545	1.0000	1.0000	0.0000	0.0000	1.0000
	1550	8.556	3.209	1.0000	0.9993	0.0000	0.0007	0.9993
	1575	9.893	1.872	1.0000	0.9694	0.0000	0.0306	0.9694
	1600	11.230	0.535	1.0000	0.7037	0.0000	0.2963	0.7037
UAL	**1610**	**11.765**	**0.000**	**1.0000**	**0.5000**	**0.0000**	**0.5000**	**0.5000**
	1625	12.567	–0.802	1.0000	0.2113	0.0000	0.7887	0.2113
Max lot avg.	**1650**	**13.904**	**–2.139**	**1.0000**	**0.0163**	**0.0000**	**0.9837**	**0.0163**
	1675	15.241	–3.476	1.0000	0.0003	0.0000	0.9997	0.0003
USL	**1700**	**16.578**	**–4.813**	**1.0000**	**0.0000**	**0.0000**	**1.0000**	**0.0000**

Note that essentially all lots with true lot averages between 1450 and 1550 will be accepted ($P_{ac} \approx 1.00$). The lot-to-lot standard deviation was estimated at 27.1. With the process on target, about 95 percent of the lots will be within ±2 standard deviations of the target or 1500 ± 54 (1446 to 1554). Nearly all product will be accepted if the process stays on target.

24A.4.4 Developing the OC Curve—1 bag/lot

An OC curve has also been developed for a sample size of 1 bag/lot. The computations are outlined as follows.

The desired P_{ac} at the upper lot average limit = 0.01
z = –2.33 (from normal tables for probability = 0.01)
The standard deviation of lot average (sample size = 1) = 34.2

$$
\begin{aligned}
\text{Lower acceptance limit (LAL)} &= \text{Lower lot average limit} - z\sigma_{AVG} \\
&= 1350 - (-2.33)(34.2) \\
&= 1350 + 79.7 \\
&= 1429.7 \text{ (Round to 1430)} \\
\text{Upper acceptance limit (UAL)} &= \text{Upper lot average limit} + z\sigma_{AVG} \\
&= 1650 + (-2.33)(34.2) \\
&= 1650 - 79.7 \\
&= 1570.3 \text{ (Round to 1570)}
\end{aligned}
$$

The probability of acceptance P_{ac} for both lower and upper acceptance limits has been computed for a range of true lot averages in the same manner as for the sample size of four bags/lot.

Table 24A.9 shows the computations for several points on the OC curve. The points for the LSL, USL, LAL, UAL, minimum and maximum acceptable lot averages, and the target are shown in bold type.

Note that lots with a true lot average at the target value of 1500 will be rejected about 4 percent of the time. About 2 percent of the time, the measured seal strength will be smaller than the lower acceptance limit of 1430, and about 2 percent of the time the measured seal strength will be greater than the upper acceptance limit of 1570. The poor discrimination is the result of the relatively large variability for measurement and bags within lot that makes it difficult to precisely determine the true lot average. Using a sample size of four bags/lot reduces the effect of both of these large variance components.

24A.4.7 The OC Curves

The OC curves for the two sample sizes are shown in Figure 24A.2. Note that both plans result in the same protection for the customer. The four bags/lot sample provides more protection for the producer than the one bag/lot sample. This is shown on the OC curves as a higher probability of accepting product made near the specification limits.

The acceptable lot average limits for seal strength for the two acceptance plans are compared in Table 24A.10. Both plans have the same target, 1500 grams. They have the same specification limits, 1300 to 1700 grams, and the same acceptable lot average limits, 1350 to 1650 grams.

Table 24A.9. Computations for OC curve—1 bag/lot.

	Lot avg	z_{lower}	z_{upper}	$P_{ac}(l)$	$P_{ac}(u)$	$P_{re}(l)$	$P_{re}(u)$	P_{ac}
LSL	**1300**	**–3.801**	**7.895**	**0.0000**	**1.0000**	**1.0000**	**0.0000**	**0.0000**
	1325	–3.070	7.164	0.0011	1.0000	0.9989	0.0000	0.0011
Min lot avg.	**1350**	**–2.339**	**6.433**	**0.0096**	**1.0000**	**0.9904**	**0.0000**	**0.0096**
	1375	–1.608	5.702	0.0539	1.0000	0.9461	0.0000	0.0539
	1400	–0.877	4.971	0.1902	1.0000	0.8098	0.0000	0.1902
	1425	–0.146	4.240	0.4419	1.0000	0.5581	0.0000	0.4419
LAL	**1430**	**0.000**	**4.094**	**0.5000**	**1.0000**	**0.5000**	**0.0000**	**0.5000**
	1450	0.585	3.509	0.7207	0.9998	0.2793	0.0002	0.7205
	1475	1.316	2.778	0.9059	0.9973	0.0941	0.0027	0.9032
Target	**1500**	**2.047**	**2.047**	**0.9796**	**0.9796**	**0.0204**	**0.0204**	**0.9592**
	1525	2.778	1.316	0.9973	0.9059	0.0027	0.0941	0.9032
	1550	3.509	0.585	0.9998	0.7207	0.0002	0.2793	0.7205
UAL	**1570**	**4.094**	**0.000**	**1.0000**	**0.5000**	**0.0000**	**0.5000**	**0.5000**
	1575	4.240	–0.146	1.0000	0.4419	0.0000	0.5581	0.4419
	1600	4.971	–0.877	1.0000	0.1902	0.0000	0.8098	0.1902
	1625	5.702	–1.608	1.0000	0.0539	0.0000	0.9461	0.0539
Max lot avg.	**1650**	**6.433**	**–2.339**	**1.0000**	**0.0096**	**0.0000**	**0.9904**	**0.0096**
	1675	7.164	–3.070	1.0000	0.0011	0.0000	0.9989	0.0011
USL	**1700**	**7.895**	**–3.801**	**1.0000**	**0.0000**	**0.0000**	**1.0000**	**0.0000**

Table 24A.10. Lot acceptance limits for seal strength.

	Seal strength (g)	
	Lower	**Upper**
Four bags/lot	1390	1610
One bag/lot	1430	1570

The operating interval between the lot acceptance limits is 220 grams for the four bags/lot sample plan compared to 140 grams for the one bag/lot sample plan. This is a 57 percent wider operating interval for the producer while maintaining the customer's risk at no more than 2.5 percent.

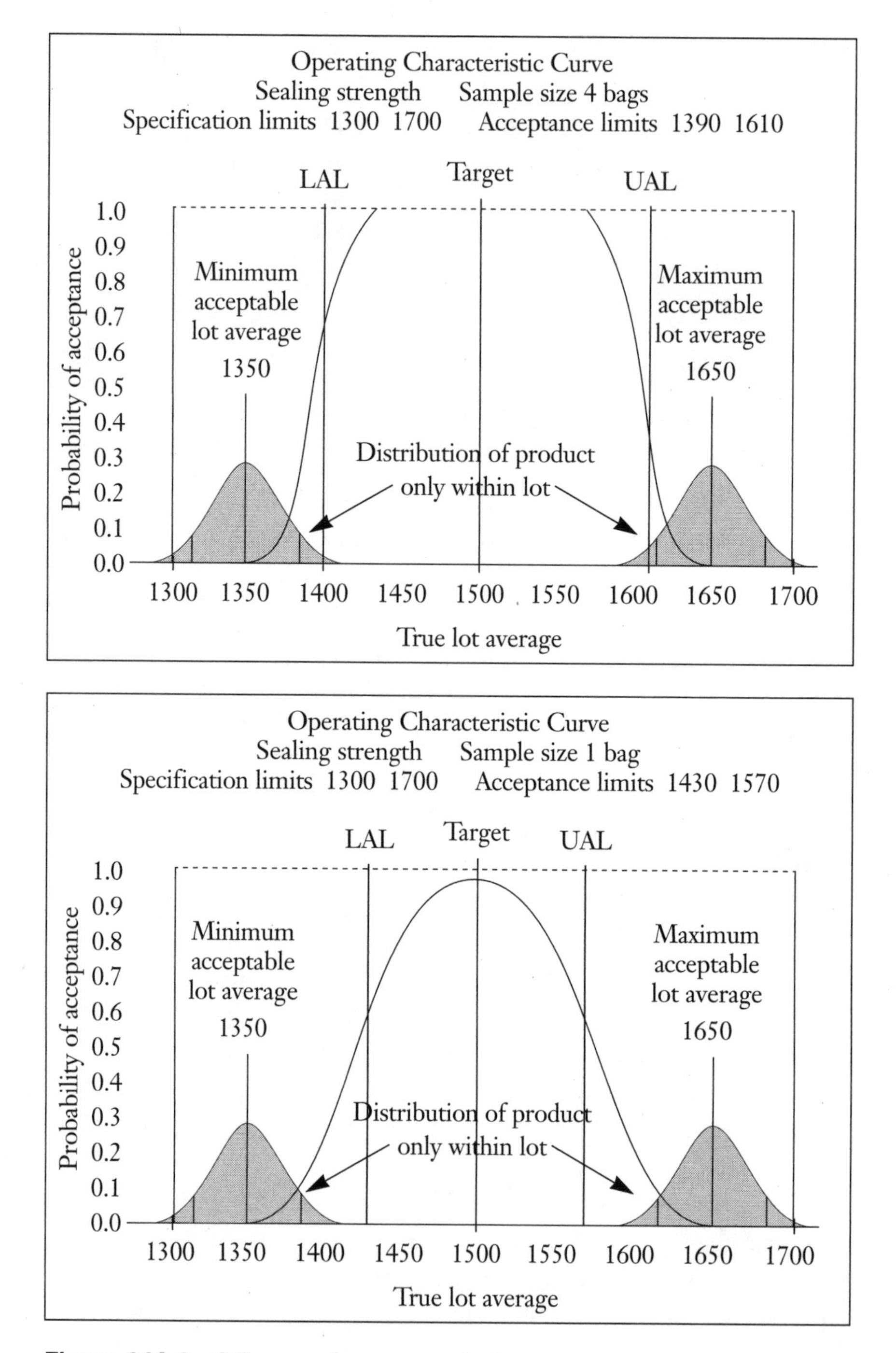

Figure 24A.2. OC curves for two sample sizes.

Glossary

The terms included here are critical to the preparation, implementation, and utilization of specifications. Wherever possible, the terms taken were from existing standards. In some cases the definitions have been modified to fit the authors' perspective on specifications.

The reference shown in brackets [...] after each entry identifies the source document and paragraph of the definition.

- [A1, x.x] denotes ANSI/ASQC A1-1987
- [A2, x.x] denotes ANSI/ASQC A2-1987
- [A3, x.x] denotes ANSI/ASQC A3-1987
- [8402, x.x] denotes ANSI/ISO/ASQC A8402-1994

The reference [CIC] marks a definition developed by the authors.

Accept (acceptance sampling sense): To decide that a batch, lot, or quantity of product, material, or service satisfies the requirement criteria based on the information obtained from the sample(s). [A2, 1.1]

Acceptance limit: The limiting value a reported lot average may take and still meet specifications. [CIC]

Attributes, method of: Measurement of quality by the method of attributes consists of noting the presence (or absence) of some characteristic or attribute in each of the units in the group under consideration, and counting how many units do (or do not) possess the quality attribute, or how many such events occur in the unit, group, or area. [A2, 1.8]

Batch: See *Lot (batch).*

Certificate of analysis: A formal report of test results, indicating that testing was done on a lot or a shipment to a customer. See chapter 22. [CIC]

Certificate of compliance: A document signed by an authorized party affirming that the supplier of a product or service has met the requirements of the relevant specification, contract, or regulation. [A3, 3.4.1]

Certificate of conformance: A document signed by an authorized party affirming that a product or service has met the requirements of the relevant specification, contract, or regulation. [A3, 3.5.1]

Characteristic: A property that helps to differentiate among items of a given sample or population. *Comment:* The differentiation may be either quantitative (by variables) or qualitative (by attributes). [A2, 1.13]

Compliance: An affirmative indication or judgment that the supplier (or producer) of a product or service has met the requirements of the relevant specification, contract, or regulation; also the state of meeting the requirements. [A3, 3.4] The definitions of *compliance* and *conformance* are almost identical, except that in contrast to *conformance, compliance* refers to the supplier (or producer).

Conformance: An affirmative indication or judgment that a product or service has met the requirements of the relevant specification, contract, or regulation; also the state of meeting the requirements. [A3, 3.5] The definitions of *compliance* and *conformance* are almost identical, except in contrast to *compliance, conformance* refers to the product or service.

CPI: Chemical and Process Industries. [CIC]

Customer: Recipient of a product from a supplier. [8402, 1.9] In the context of this book the producer is the supplier and the customer is an organization rather than an individual.

Customer's (consumer's) risk: The probability of acceptance of a lot, the quality of which has a designated numerical value representing a level that it is seldom desired to accept. [A2, 1.48]

Grade: An indicator of category or rank related to features or characteristics that cover different sets of needs for products or services intended for the same functional use. *Note:* Grade reflects a planned difference in requirements or, if not planned, a recognized difference. The emphasis is on functional use and cost relationship. (See chapter 17.) [A3 2.4]

Heart cut: The material in the middle group of a sorting process. [CIC]

In-control process: A process in which the statistical measure(s) being evaluated are in a "state of statistical control." [A1, 12.16]

Inspection: Activities, such as measuring, examining, testing, or gauging one or more characteristics of a product or service, and comparing these with specified requirements to determine conformity. [A2, 1.17]

(Inspection) lot: A collection of similar units, or a specific quantity of similar material offered for inspection and subject to a decision with respect to acceptance. [A2, 1.25]

Lot (batch): A definite quantity of some product accumulated under conditions that are considered uniform. [A2, 1.30] A batch is often smaller than a lot.

Lot average: The arithmetic mean of all test results for a characteristic of interest in a lot. [CIC]

Lot size (*N*): The number of units in a lot. [A2, 1.33] This quantity may be expressed as a count, weight, mass, volume, time interval, or in some other quantitative fashion. [CIC]

Observed value: The particular value of a characteristic determined as a result of a test or measurement. [A1, 12.19]

Operating characteristic curve (OC curve) (acceptance control chart usage): A curve showing, for a given acceptance control chart configuration, the probability of accepting a process as a function of the quality level of the process. (See chapters 12 and 13.) [A1, 12.20]

Performance test: A scaled version of an actual production process application. See chapter 9. [CIC]

Population: The totality of items under consideration. *Comment:* The items may be units or measurements, and the population may be real or conceptual. Thus, population may refer to all the items actually produced in a given day or all that may be produced if the process were to continue in control. [A1, 12.21]

Procedure: A specified way to perform an activity. (8402, 1.3)

Process: A set of interrelated resources and activities which transform inputs into outputs. [8402, 1.2]

Process capability: A statistical measure of the inherent process variability for a given characteristic. (See chapter 15.) [A1, 15.3]

Producer: An organization that manufactures products. See *Supplier.* [CIC]

Producer's risk: The probability of not accepting a lot the quality of which has a designated numerical value representing a level that it is generally desirable to accept. [A2, 1.49]

Product: The result of activities or processes. *Note:* A product may include service, hardware, processed materials, software, or a combination thereof. [8402, 1.4]

Reject (acceptance sampling sense): To decide that a batch, lot, or quantity of product, material, or service has not been shown to satisfy the requirement criteria based on the information obtained from the sample(s). [A2, 1.44]

Requirements for quality: Expression of the needs or their translation into a set of quantitatively stated requirements for characteristics of an entity to enable its realization and examination. [8402, 2.3]

Resample: An analysis of an additional sample from a lot. [CIC]

Retest: An additional analysis of a sample previously analyzed. [CIC]

Rework: Action taken on a nonconforming product so that it will fulfill the specified requirements. [8402, 4.19]

s: The sample standard deviation. (See chapter 10.)

σ: The population standard deviation. See chapter 10.

Sample (acceptance sampling sense): One or more units of product (or a quantity of material) drawn from a specific lot or process for purposes of inspection to provide information that may be used as a basis for making a decision concerning acceptance of that lot or process. [A2 1.50]

Sampling: The process of selecting a sample. [CIC]

Sorting: Classification of product into categories on the basis of the observed value of a characteristic. (See chapter 17.) [CIC]

Specification: The document that prescribes the requirements with which the product or service has to conform. [A3, 3.22] *Comment:* The concept of specification has been extended also to apply to processes in this book.

Specification limits: Limits that define the conformance boundaries for an individual unit of a manufacturing or service operation. See *Tolerance limits*. [CIC]

State of statistical control: A process is considered to be in a "state of statistical control" if the variations among the observed sampling results from it can be attributed to a constant system of chance causes. [A1, 12.27]

Statistical process control (SPC): The application of statistical techniques to the control of processes. *Note:* Sometimes SPC and SQC are used synonymously. Sometimes SPC is considered a subset of SQC concentrating on tools associated with process aspects but not product acceptance measures. [A3, 3.23]

Statistical quality control: The application of statistical techniques to the control of quality. *Note:* These techniques include the use of frequency distributions, measures of central tendency and dispersion, control charts, acceptance sampling, regression analysis, tests of significance, and so on. [A3, 3.24]

Supplier: The organization that supplies product to the customer. [8402, 1.10] In the context of this book, the supplier provides inputs (goods and services) to the producer and the producer supplies product to the customer.

Supply chain: The linkage of supplier → producer → customer. [CIC]

Target: Agreed-upon value within the specification limits that produces an optimal result for the customer and/or supplier. [CIC]

Test method: A formal procedure for carrying out a chemical or physical analysis. [CIC]

Tolerance (specification sense): The total allowable variation around a level or state (upper limit minus lower limit), or the maximum acceptable excursion of a characteristic. [A2, 1.57]

Tolerance limits (specification limits): Limits that define the conformance boundaries for an individual unit of a manufacturing or service operation. *Comment:* Limits may be established either with or without the use of probability considerations. Tolerance limits may be in the form of a single limit (upper or lower) or double limits (upper and lower). Double, or two-sided limits, occur more frequently. Double limits are often stated as a symmetrical deviation from a stated value, but they need not be symmetrical.

Frequently the term *specification limits* is used instead of *tolerance limits.* While tolerance limits is generally preferred in terms of evaluating the manufacturing or service requirements, specification limits may be more appropriate for categorizing materials, products, or services in terms of their stated requirements.

For large units of output such as a piece of fabric or a heat of steel, the tolerance limits will apply to specimens from these units and conformance of a unit will be judged on the results obtained from tests of the specimens. [A2, 1.58]

Waiver: A written authorization to use or release a product that does not conform to the specified requirements. [8402, 4.17]

References

ANSI/ASQC A1-1987. *Definitions, symbols, formulas, and tables for control charts.* Milwaukee: American Society for Quality Control.

ANSI/ASQC A2-1987. *Terms, symbols, and definitions for acceptance sampling.* Milwaukee: American Society for Quality Control.

ANSI/ASQC A3-1987. *Quality systems terminology.* Milwaukee: American Society for Quality Control.

ANSI/ASQC Q3-1988. *Sampling procedures and tables for inspection of isolated lots by attributes.* Milwaukee: American Society for Quality Control.

ANSI/ASQC S1-1987. *An attribute skip-lot sampling program.* Milwaukee: American Society for Quality Control.

ANSI/ASQC Z1.4-1993. *Sampling procedures and tables for inspection by attributes.* Milwaukee: American Society for Quality Control.

ANSI/ASQC Z1.9-1987. *Sampling procedures and tables for variables for percent nonconforming.* Milwaukee: American Society for Quality Control.

ANSI/ISO/ASQC A8402-1994. *Quality management and quality assurance—Vocabulary.* Milwaukee: American Society for Quality Control.

ANSI/ISO/ASQC Q9000-1-1994. *Quality management and quality assurance standards—Guidelines for selection and use.* Milwaukee: American Society for Quality Control.

ANSI/ISO/ASQC Q9001-1994. *Quality systems—Model for quality assurance in design, development, production, installation, and servicing.* Milwaukee: American Society for Quality Control.

ANSI/ISO/ASQC Q9002-1994. *Quality systems—Model for quality assurance in production, installation, and servicing.* Milwaukee: American Society for Quality Control.

ANSI/ISO/ASQC Q9003-1994. *Quality systems—Model for quality assurance in final inspection and test.* Milwaukee: American Society for Quality Control.

ANSI/ISO/ASQC Q9004-1994. *Quality management and quality systems—Guidelines.* Milwaukee: American Society for Quality Control.

ASQC Chemical and Process Industries Division, Chemical Interest Committee. 1987. *Quality assurance for the chemical and process industries—A manual of good practices.* Milwaukee: ASQC Quality Press.

———. 1992. *ANSI/ASQC Q90/ISO 9000 Guidelines for use by the chemical and process industries.* Milwaukee: ASQC Quality Press.

Bowker, A. H., and G. J. Lieberman. 1972. *Engineering statistics.* 2d ed. Englewood Cliffs, N.J.: Prentice Hall.

Box, G. E. P., W. H. Hunter, and J. S. Hunter. 1978. *Statistics for experimenters.* New York: John Wiley & Sons.

BSR-ASQC Z1.10. Forthcoming. *Standard method for calculating process capability and performance measures.* Milwaukee: American Society for Quality Control.

Chrysler Corporation, Ford Motor Company, and General Motors Corporation. *Fundamental statistical process control reference manual.* Southfield, Mich.: Automotive Industry Action Group.

Duncan, A. J. 1986. *Quality control and industrial statistics.* 5th ed. Homewood, Ill.: Irwin.

Grubbs, F. E., and H. J. Coon. 1954. On setting test limits relative to specification limits. *Industrial Quality Control* 10, no. 3:15–20.

Guttman, I., S. S. Wilks, and J. S. Hunter. 1965. *Introductory engineering statistics.* 2d ed. New York: John Wiley & Sons.

Herman, J. T. 1989. Capability index—Enough for the process industries? In *43rd annual quality congress transactions.* Milwaukee: American Society for Quality Control.

Hicks, C. R. 1956. Fundamentals of analysis of variance. Parts 1–3. *Industrial Quality Control* (August): 17–20; (September): 5–8; (October): 13–16.

Juran, J. M. 1974. *Quality control handbook.* 3d ed. New York: McGraw-Hill.

Kackar, R. N. 1985. Off-line quality control, parameter design and the Taguchi method. *Journal of Quality Technology* 17, no. 4:176–188.

Kane, V. E. 1986. Process capability indices. *Journal of Quality Technology* 18, no. 1:41–52.

Kittlitz, R. G. 1987. *PPK distribution.* Seaford, Del.: DuPont Report.

Natrella, M. G. 1963. *Experimental statistics—Handbook 91.* Washington, D.C.: National Institute of Standards and Technology.

Nelson, L. S. 1995. Using nested designs: I. Estimates of standard deviation. *Journal of Quality Technology* 27, no. 2:169–171.

Occupational Safety and Health Administration (OSHA). Process safety management of highly hazardous chemicals standard, title 29. Code of Federal Regulations, part 1910.119. *Federal Register* 57, no. 36 (24 February 1992): 6356–6417.

Ryan, T. P. 1989. *Statistical methods for quality improvement.* New York: John Wiley & Sons.

Schilling, E. G. 1982. *Acceptance sampling in quality control.* New York: Marcel Dekker.

Taylor, J. K. 1987. *Quality assurance of chemical measurements.* Chelsea, Mich.: Lewis Publishers.

Taylor, W. A. 1992. *Guide to acceptance sampling.* Lake Villa, Ill.: Taylor Enterprises.

Wheeler, D. J., and R. W. Lyday. 1989. *Evaluating the measurement process.* 2d ed. Knoxville, Tenn.: SPC Press.

Index